D1084603

SCIENCE IN FRANCE IN THE REVOLUTIONARY ERA

This is Number 7 in a series of monographs in the history of technology and culture published jointly by the Society for the History of Technology and The M.I.T. Press. The members of the editorial board for the Society for the History of Technology Monograph Series are Melvin Kranzberg, Cyril S. Smith, and R. J. Forbes.

Publications in the series include:

History of the Lathe to 1850, Robert S. Woodbury

English Land Measuring to 1800: Instruments and Practices, A. W. Richeson

The Development of Technical Education in France 1500–1850, Frederick B. Artz

Sources for the History of the Science of Steel 1532–1786, Cyril Stanley Smith

Bibliography of the History of Technology, Eugene S. Ferguson

The Hollingworth Letters: Technical Change in the Textile Industry, 1826–1837, Thomas W. Leavitt, editor

Science in France in the Revolutionary Era, Described by Thomas Bugge, Danish Astronomer Royal and Member of the International Commission on the Metric System (1798–1799), Maurice P. Crosland, editor

SCIENCE IN FRANCE IN THE REVOLUTIONARY ERA

Described by Thomas Bugge,
Danish Astronomer Royal and Member
of the International Commission
on the Metric System (1798–1799)

Edited with Introduction and Commentary by
Maurice P. Crosland

With Extracts from Other Contemporary Works

Published Jointly by
The Society for the History of Technology
and THE M.I.T. PRESS
Cambridge, Massachusetts, and London, England

Photoset in Lumitype Latine and Optima
by St Paul's Press Ltd, Malta.
Printed and bound in the United States of America
by The Colonial Press, Inc.

SBN 262 03029 2 (hardcover)

Library of Congress catalog card number: 79–86611

CONTENTS

CONTENTS OF ORIGINAL DANISH EDITION

Parts included in present edition are shown with an asterisk.
Page numbers are shown to indicate comparative lengths of chapters.

PREFACE

This book provides a description of the scientific scene and some aspects of the cultural scene in France in the Revolutionary era, a period when some could claim that science really came of age. It was a time when science in France blossomed to its full maturity, moving from a stage in which it had been largely of interest to a handful of serious students to a situation where it became the everyday concern for a significant body of professionals. The author of this account, Thomas Bugge, was a Danish astronomer who systematically recorded places, events, and people in Paris during his six-month stay in 1798–1799 to attend what might be described as one of the first international scientific conferences.

Although it is hoped that it will be of some use to those interested in the history of science and of education, no claim can be made for this book beyond that of being a contribution to the literature of the history of scientific institutions. It is, therefore, not a history of science but a source for historians of science and others. When the definitive history of science in France after the Revolution is written, it will be based on a wide variety of evidence, ranging from debates in the assemblies and personal evidence in letters to technical accounts of scientific research. All will contribute to an assessment of the inter-relation between science and society in this crucial period. When this tribunal comes to consider what evidence is relevant, I should like to enter a plea for consideration of the accounts provided by visitors to France with some special interest in science.

The description of events and people by an eyewitness provides a dimension lacking from more formal records such as the minutes of the National Institute. The critical use of such personal testimony can complement official documents to provide a more complete picture. As explained in the Introduction, Bugge was in a particularly good position to assess the scientific scene. Yet the impressions of visitors other than men of science cannot be ignored. It was, after all, the favorable impact of the Ecole Polytechnique outside France which led to the establishment of technological universities in other countries. The German *technische Hochschulen* were modeled directly on the Ecole Polytechnique and sprang up in the wake of its growing reputation: Prague (1806), Vienna (1815), Karlsruhe (1825), Munich (1827), Dresden (1828), Stuttgart (1829), and Hanover (1831). Polytechnics were founded in Europe from Copenhagen to Zurich and even at St. Petersburg. American education too was affected. One of the graduates of the Polytechnique, Crozet, brought its methods to the U.S. Military Academy at West Point, and Greene introduced the educational principles of the Polytechnique at Rensselaer Polytechnic Institute (Troy, New York). The Massachusetts Institute of Technology (founded in 1865) derives from the same tradition.

An admirable survey of eighteenth-century institutions concerned with the teaching of science in France has been published recently.[1] Unfortunately, it stops short of the complex Revolutionary period. The present book may go part of the way toward filling this important gap. The editor offers this book as a contribution from a particular viewpoint to a greater knowledge and understanding of the scientific institutions established in France after the Revolution. Just as it was believed that society could be reformed by the establishment of appropriate political institutions, so scientific institutions had been set up as

[1] R. Taton, ed., *Enseignement et diffusion des sciences en France au XVIIIe siècle*, Paris, 1964.

the basis of advance in science. The evidence of Bugge is supplemented by that of several other visitors to Paris at the time of the Directory or Consulate where the interests or education of these visitors prompted them to comment on the state of institutionalized science. The bibliography contains a selection of such published accounts, many of which have been completely overlooked hitherto.

Although Thomas Bugge wrote in Danish, contemporary translations of the account of his visit to France were published in English and German.[2] The English edition of 1801 is reproduced here with corrections and occasional omissions and alterations in the interests of accuracy, brevity, and intelligibility. Some minor changes in spelling, capitalization, and punctuation have been made. Certain archaic words have been changed, particularly if they might mislead the reader.[3] Bugge, in accordance with current French practice, gives dates according to the Republican calendar, the details of which are explained here in a glossary. In this edition the dates according to the Republican calendar are given in a standardized form and the equivalent on the ordinary calendar is usually added. A very large number of proper names were incorrectly given in the 1801 edition. This has been rectified, at least in the great majority of cases. The occasional use of the title C[itizen] is abandoned for the sake of consistency.

Several parts of the book (Chapters 5 and 7 and part of Chapter 4) now appear for the first time in English. The original Danish edition contained thirty-one "letters" or

[2]See bibliography, pp. 216–219.

[3]Thus arc is substituted for "arch" in geodesic measurements, cellars for "caves," factory for "manufactory," morning for "forenoon," platinum for "platina," prize for "premium," program for "programma," room or gallery for "saloon," and school for "nursery" and "seminary." Other changes are indicated by footnotes in the text, marked "J. J." (John Jones, the translator of the 1801 English edition).

In one place, the terms "complementary" and "noncomplementary" are used in the 1801 edition to refer to the odd and even days of the republican week of ten days. The proper use of the term "complementary day" is explained in the glossary under "Calendar."

chapters. Of these, only a selection of the first eighteen was translated into English, although the translator gave the impression that this edition was complete. A complete German translation was made in 1801 with the assistance of Bugge himself and has been used as the source of the additional material, the Danish edition being consulted as a check. Details of the French budget for science and art, scientific societies, instrument makers, the metric system, steam engines, and the manufacture of gunpowder and cannon have been included. In order that the reader may see what selection has been made, a full list of contents of the original edition is given, indicating comparative lengths of chapters and showing what sections have been omitted.

The principle adopted in editing this work has been to preserve the bias toward the description of scientific institutions without suppressing the author's comments on such important centers of culture as the Louvre and the Bibliothèque Nationale. It would have been possible to edit Bugge's work so that only the best in the French scene was included, but I have avoided this one-sided approach and indeed the book opens with an account of the primary schools—almost a national scandal. The Observatory too was not at its best immediately after the Revolution. In contrast, there are full descriptions of such famous institutions as the Ecole Polytechnique, the Muséum d'Histoire Naturelle, and the First Class of the Institute. Notes to the text have been kept to an absolute minimum. Most of Bugge's account speaks for itself. To have elaborated every allusion would have doubled the size of the book without increasing its value proportionately. Bugge's footnotes are marked by symbols whereas the editor's are numbered consecutively throughout each chapter.

The price to be paid for an over-all survey is lack of thoroughness in the treatment of any one aspect. In many cases this text will serve to raise certain questions rather than to provide final answers. For example, the summary treatment of the metric system may annoy those who find this of particular interest. It will be justified, however, if it stimulates a complete study of the establishment of the metric system, including its social and political implications.

In the preparation of this book I have received valuable assistance from several friends and the staffs of various libraries. Dr. W. A. Smeaton generously agreed to read the original typescript and was able to make some helpful suggestions and to correct many errors. Professor Roger Hahn has not only provided me with several important references, but also made a considerable number of detailed and useful criticisms, for which I am grateful. Dr. Eric Forbes was kind enough to read the chapter on the metric system and to help with the translation of several difficult passages. Among others, whose help with the historical background I gratefully acknowledge are Professor John Bosher and Dr. Ernst Wangermann. I would like especially to thank my wife for her invaluable assistance in translating many passages from the German. Although I take this opportunity of expressing my gratitude to all concerned, responsibility for the presentation of this work must rest with me. I cannot be certain about the complete absence of mistakes or misprints in this edition, but any that remain can be no more than a small fraction of those in the original English edition.

I should like to thank the staff of the Brotherton Library, University of Leeds, for obtaining for me on interlibrary loan a copy of the original English edition of this work from Bristol City Library, a copy of the German edition, and a microfilm of the Danish edition from the Kongelige Bibliotek, Copenhagen. Other material was obtained in the library of the British Museum. I must thank MM. les Secrétaires Perpetuels of the Académie des sciences for allowing me to consult the dossier on Thomas Bugge in the archives of the Académie. The quotation from a manuscript in the rare book room of the library of the University of California at Berkeley is by kind permission of the Department of Rare Books and Special Collections.

Finally I should like to thank the secretarial staff of the Department of Philosophy, University of Leeds, for the typing of this manuscript.

MAURICE P. CROSLAND

University of Leeds
September 1968

SCIENCE IN FRANCE IN THE REVOLUTIONARY ERA

INTRODUCTION

France in the eighteenth century was a major world power. Under Louis XIV (1643–1715), France had asserted her place as a dominant European state and within France the monarchy was supreme. The next king, Louis XV, should have been more aware of the disasters that threatened his country from without and within. The extravagant life at court could not hide the many social injustices in France. Despite censorship, Voltaire, the encyclopedists, and Rousseau made powerful criticisms of the authority of state and church and in their different ways looked forward to a more enlightened society. Matters came to a head in the reign of Louis XVI. To the general feeling of unrest was added an inescapable financial crisis in the 1780's. What is usually referred to as the French Revolution came as a complex succession of revolutions with diverse aims and at different levels in society. In 1787, the nobles rebelled against the king. By 1789, the professional classes were playing a leading role in demands for reform of the many abuses of the *ancien régime*. Later the Paris mob was to play a crucial part in the final overthrow of the monarchy. We may look to the widespread poverty of the masses for the economic origins of the Revolution.

In May 1789, the king took a step which was to weaken his position as an absolute monarch. He summoned the States General, a specially elected assembly of clergy, nobility, and bourgeoisie. It was the latter, the Third Estate, which eventually took control and proclaimed itself the true representative assembly of the French people. This "National Assembly" began by abolishing feudal privileges and it proclaimed the Declaration of the Rights of Man. Its decree, which confiscated church lands, was passed as a means of solving the ever-present financial problems of the state but later there were prominent antireligious elements in the Revolution. This period marks the beginning of the secularization of French society and institutions. For two years there was a constitutional monarchy but the outbreak of war created a situation in which extremists were able to strip the king of all his powers and finally bring about his execution. In September 1792, France was proclaimed a republic and was ruled by a succession of dominant personalities in the revolutionary assembly known as the Convention.

The new France appeared to be threatened by the other European powers and an army was raised. After initial defeats, the scales were turned by general mobilization in August 1793 and by the massive organization of supplies. The victorious Republican army, of which the most famous commander was to be Napoleon Bonaparte, began to push back the former frontiers of France. Neighboring states were proclaimed republics with the aid of nationals sympathetic to the ideals of the Revolution.

A further threat of foreign invasion in the summer of 1793 as well as the activities of counterrevolutionaries produced the violent period of the Terror. Thousands of people considered hostile to the Revolution were arrested as suspects and many were executed. They included not only obvious figures like Marie Antoinette but also men who had helped to found the Republic. The great chemist Lavoisier was among those executed in Paris in May 1794. Although it was his activities as a tax farmer and not his scientific work that was under attack, it is symptomatic of the mood of the time that even his eminent scientific reputation and his work on government committees were not considered valid reasons for sparing him. With the execution of Robespierre in July 1794, this violent and negative phase of the Revolution came to an end and a constructive phase slowly began. New educational and scientific institutions which had been planned earlier by the Committee of Public Instruction of the Convention were now set up. Most of these continued to flourish under the Directory, which ruled France from November 1795 until the coup d'état of Napoleon Bonaparte in November 1799.

The government of the Directory gave political power back to the bourgeoisie. It outlawed the Jacobins, who had been the dominant group in the Convention, but it feared the Royalists equally. The two assemblies of this period, the Council of Five Hundred and the Council of the Ancients, appointed five directors who were to form the government. The constitution was unwieldy and inefficient and several of the directors were corrupt. Yet, despite its faults, the Directory managed to perpetuate many of the ideals and institutions of the Revolutionary Convention while avoiding some of its worst extremes.

This, then, is the general political background of Bugge's account, a background woven from a complex pattern of revolution and the threat of counterrevolution. Although in the Revolutionary period the scene was a continually changing one, the period of Bugge's visit, 1798–1799, was relatively stable. However, since April 1792, France had been continually at war and when the Danish visitor expressed his desire for an end to the war, his sentiments must have been shared by many Frenchmen.

At the time of Bugge's visit, French scientists would with some justification have considered Paris the center of the scientific world. Indeed the purpose of the visit of the Danish astronomer depended to a considerable extent on the prestige of French science as well as the record of success of the French armies. In the 1780's under the *ancien régime*, Lavoisier had constructed the science of chemistry on completely new foundations. The prowess of the French in several other sciences and particularly mathematics was hardly less remarkable. The Revolution gave science a new impetus and perhaps a new direction. It became more democratic in its recruitment. The new educational system was open to even the humblest student, who could obtain financial support from the state to attend one of the advanced schools in Paris if he could demonstrate his ability. The educational system gave prominence to science on both ideological and utilitarian grounds and France's leading scientists were recruited as professors in the new establishments for higher education.

As a university teacher himself, Bugge was able to give an informed critical assessment of the organization, staffing, and curriculum of the new French educational system. He visited all the major institutions in Paris. His account throws new light on many of them, including the First Class of the Institute, the official body of French science. As an astronomer, Bugge naturally visited the National Observatory with particular interest and the very full account that he gives makes a further contribution to the history of this institution. He provides a record of his conversations with French scientists including the veteran astronomer Cassini and the great crystallographer Haüy. His collaboration with his French colleagues in the expanded Commission of Weights and Measures enabled him to see the problems of new metric standards from the inside. Chapter VII includes several criticisms by Bugge of the metric system. Having an interest in all aspects of science, Bugge did not neglect even the instrument makers and his observations (omitted from the original English edition) provide information lacking in the standard histories.[1] His visit took place in the midst of the Revolutionary Wars but he was not prevented from giving an eyewitness account of the manufacture of munitions. He was one of the very few foreigners who was able to inspect the exhibition of French

[1]For example, M. Daumas, *Les instruments scientifiques au XVIIe et XVIIIe siècles*, Paris, 1953.

industry, held in Paris in 1798—the first such exhibition ever to be held. In the Danish visitor's account of large-scale government financial support for science, we see in embryo the present-day state of affairs.

In the Revolutionary and Napoleonic periods, memoirs of travelers in France streamed from the presses. Any literate person who had spent even a few days in the country which had aroused first the horror and then the curiosity of neighboring countries could become an "author" overnight. There were so many tracts written by visitors to Paris after the French Revolution that one of these pocket-sized books could understandably bear the title *Yet Another Description of Paris* (*Encore un tableau de Paris*). The theaters, the opera, public buildings, all were visited and impressions recorded. Politics and religion were also popular subjects for discussion. The world of fashion was not forgotten, but any reader whose principal interest is the wild attire of the young men of the Directory period, the *incroyables*, should read no further. The author of this account was not concerned with the gay life of the French capital. He had come to Paris as a man of science and it was the scientific and cultural life which he was interested to observe and record. No other contemporary account pays so much attention to the *scientific* life of Paris.

The position of the Danish astronomer Thomas Bugge (1740–1815) was an exceptional one, for his mission to Paris in 1798 was as a scientific representative of a neutral state to take part in the final stages of the establishment of the metric system. The committee which Bugge joined in Paris was one of the first international scientific committees, with representatives from the Netherlands, Switzerland, Denmark, the Italian states, and Spain, as well as France. Few people, even historians of science, have heard of Thomas Bugge (or Bÿggé),[2] who, as director of the Royal

[2]His name is spelled this way in the English translation of his *Travels* published in 1801.

Observatory in Copenhagen, was the successor of Tycho Brahe and Rømer. Before concerning ourselves with Bugge's life, however, it would be appropriate to say something about the French scene. The reader interested in the history of scientific and educational institutions deserves an explanation of why this particular account (previously generally unknown) of a visit to France has been chosen for republication.

The Danish scientist was visiting a country unique not only in its revolution but even more in the institutions which had been established after the overthrow of the monarchy. It is not with successive political institutions that we are concerned here, however, but with the scientific establishments. These reflected in the eyes of the Convention the value of science as a servant of the state and as a means of enlightenment. These sentiments, combined with the importance attached to education, had the effect of making Paris a world center for higher education in science. One contemporary explanation of this looked to the prominent part played by men of science in overcoming the crisis of 1793–1794 when France was threatened by invasion of foreign armies. The supply of gunpowder and the construction of cannon was put in the hands of men like Monge, Fourcroy, Guyton de Morveau, and Berthollet, and their work resulted in an immediate and spectacular increase in army supplies, thus averting the crisis. This conclusion followed:

> The *savans* who had effected such great things, for some time enjoyed unlimited influence. It was well known that to them the Republic was indebted for its safety and very existence. They availed themselves of this favourable moment for insuring to France that superiority of knowledge which had caused her to triumph over her enemies.[3]

[3][F. W. Blagdon], *Paris as it was and as it is*, London, 1803, II, 277.

It is certainly significant that the same four men were also prominent in the work of the Committee of Public Instruction, which made recommendations for setting up a new educational structure in France. This provides an unparalleled example of scientists as legislators "writing their own ticket."

The value of the Danish visitor's account may be considered under six heads:

1. Bugge was in an almost unique position as a foreign man of science invited to postrevolutionary France in time of war and, fortunately, he recorded his impressions. The majority of accounts of life in France had been written by British visitors. There were a few foreign eyewitnesses of the early stages of the Revolution, but war with England was declared on 1 February 1793 and on 18 October it was decreed that any Englishman remaining in France would be arrested and his goods confiscated. This was the state of affairs until the Peace of Amiens in 1802. The most famous case of a foreign visiting scientist in France in the period 1789–1815 is that of Sir Humphry Davy. Many erroneous statements have been made about this visit. In brief, however, Davy received permission to travel *through* France to Italy in 1813. He stayed for two months in Paris and for one month in Montpellier but apparently he kept no record of his stay in France.[4]

2. The author of this account was primarily interested in the scientific institutions of the French capital and was able to give a reasonably objective description of them.

The objective description of the scientific scene in Paris by Bugge may be appreciated all the more if compared with other accounts such as that by Henry Redhead Yorke in his *Letters from France*. Yorke claimed an interest in science and considered that France's scientific achievements were "without a rival."[5] Yet he described Paris as

[4]John Davy, *Memoirs of the life of Sir Humphry Davy*, London, 1836, I, 467.

[5]H. R. Yorke, *Letters from France*, London, 1814, II, 336.

"the most polluted den in the civilized world" and the Parisian *savans* as "in general a gang of the vilest ruffians in the whole world."[6] He announced his intention of devoting a further book for the British public "to cure them, if possible, of their mania for visiting this diabolical metropolis."[7] Yorke, once a fierce protagonist of the Revolution, had become disillusioned and had swung to the opposite extreme.

3. Bugge was not a casual visitor, interested primarily in spectacle. His stay of six months in Paris gave him an opportunity to obtain rather more than the superficial view of the tourist.

Several interesting comments on the scientific scene in Paris were made by J. F. Reichardt, who was there in the winter of 1802–1803, but what this young German appreciated most were the theater and the opera. He complained that attendance at a solemn reopening session of the Collège de France deprived him of his usual nightly entertainment.[8] Reichardt also had the habit of describing any Frenchman who really impressed him (for example, Chaptal, Vauquelin) as not typically French but rather Germanic!

4. Bugge was in an official position. He therefore had access to people and meetings not open to the ordinary visitor. In addition, he was not afraid to explore on his own, as his visit to the National Observatory shows.

5. The description is a contemporary one. It is therefore more reliable than the numerous French social memoirs written after the Bourbon restoration which provide reminiscences of the earlier period.

6. The year of Bugge's visit was one of comparative stability. Most of the scientific institutions created three years earlier had been functioning long enough to overcome their initial problems. On the other hand, this is a

[6] *Ibid.*, pp. 48 and 39.
[7] *Ibid.*, p. 59.
[8] J. F. Reichardt, *Un hiver à Paris*, Paris, 1896, p. 34.

genuine picture of *Revolutionary* and not Napoleonic France. We have the *écoles centrales* and not the *lycées* and in 1798 the Ecole Polytechnique was still six years away from the military regime imposed by Napoleon Bonaparte in 1804.

Granted many of the advantages, however, no one would wish to claim for Bugge's *Travels in the French Republic* that it is a great work of literature. The style which emerges in the English translation is pedestrian and the author often tends to catalogue information.[9] Some parts of the original, which consist of no more than a list (for instance, the ninety-seven *écoles centrales* proposed for all the departments) have been omitted or at least drastically abbreviated. In other cases, as where the author describes the first industrial exhibition held in France, it is of some value to have a survey of the exhibits together with an estimate of their worth.

Although Bugge's account is reasonably objective, the text not unnaturally reveals certain predilections of the author. He was impressed by the institutions resulting from the legislation of the Convention and the Directory, but he was not happy about some of the earlier history of the Republic. He makes it quite clear how opposed he was to the violence of the Terror. He condemns the Jacobins outright and sympathizes with the royalist astronomer Cassini. He condemns unequivocally the spirit of iconoclasm which affected science and the arts in 1793–1794. On the other hand, he seems to have appreciated that the equipment of the new schools and public collections depended largely on the seizure of crown property and confiscations from those former nobles who had fled the country. Indifferent to the new French ideal of republicanism, Bugge was a conservative in politics. Yet, while many of his contemporaries could find nothing but blame

[9]The translation was made by John Jones, LL.D., of Yarmouth. He was assisted by Dr. William Dickson, author of works on the abolition of slavery.

to attach to the First Republic, Bugge's deep concern for science made him favorably disposed to the French government as a patron of science.

What impressed nearly everyone who was able to compare the France of the *ancien régime* with the France of the Directory and Consulate was the prominence given to science after the Revolution. To some this appeared to be at the expense of literature:

> I must . . . acknowledge that literature, which formerly held the first degree in the scale of the moral riches of this nation, is likely to decline in priority and influence. The sciences have claimed and obtained in the public mind a superiority resulting from the very nature of their object; I mean utility. The title of *savant* is not more brilliant than formerly, but it is more imposing; it leads to consequence, to superior employments and above all, to riches.[10]

The view of science as connected with wealth and position in society is one which belongs more to the Napoleonic period, but the idea of a career structure in science was present earlier and indeed was embodied in the scientific institutions established by the Convention in 1794–1795.

The general interest in science in France impressed many visitors. J. G. Heinzmann, who came to Paris in 1798, reported the opinion of the veteran astronomer Lalande:

> Our young men are beginning to study the abstract sciences by way of preference. In the fifty years that I have been teaching here and have observed the state of education, I have never seen so many students of mathematics as at present. And this partiality for real science [*les sciences solides*] seems to become more general every day.[11]

[10] [Blagdon], *Paris as it was*, I, 395.
[11] [J. G. Heinzmann], *Voyage d'un Allemand à Paris*, Lausanne, 1800, p. 117.

Among reasons for the great interest in mathematics was the dominant place which the Ecole Polytechnique came to have in higher education. Entry was by means of a competitive examination in mathematics taken by candidates from all over France. Besides the direct effect of ensuring that all students entering the Ecole were proficient in mathematics, it also had a far-reaching effect outside the walls of the Ecole:

> The Ecole polytechnique promotes the study of mathematics amongst a considerable number of young men who aspire to be admitted as scholars; those who succeed are thus placed in a career of promotion and the unsuccessful candidates have acquired a portion of mathematical knowledge which may be of use to them in other situations of life.[12]

Some visitors were particularly impressed by the number of scientific courses of various kinds available in Paris; others commented on the ability of the lecturers either as eloquent exponents of their subject or as leading men in their particular field. Public courses, which were usually open to all without payment, greatly impressed foreign students. The account given by Heinzmann, although exaggerated slightly, illustrates this impression on the foreign observer:

> Professors, teachers, public functionaries have continued in the exercise of their duties without wishing for the least salary. Young men can have daily public lectures on physics, natural history, mathematics and architecture, free and from able men. One even sees posters on the walls offering private lessons at stated times and these are open to everyone without any fee being required.[13]

[12] [Anon.], *Some account of . . . the physical sciences in Paris*, [1809], MS., Rare Book Dept., University of California, Berkeley, p. 11.
[13] Heinzmann, *Voyage d'un Allemand*, p. 119.

The Englishman Yorke too was impressed by the fact that the financial support for the institutions for higher education came not from the individual students but from the budget of the Minister of the Interior:

> All these different establishments, which honour a nation, are supported entirely at the expense of the state; the professors are paid out of the public revenues, and students of all ages and countries are at liberty to consult their libraries, and attend their lectures free of any expense. [Such] institutions . . . illustrate the magnificence and power of a nation . . .; they invite the inquisitive traveller to prolong his residence in the capital, and they allure the foreign student to crown his labours by an attendance at those public lessons, which are delivered gratuitously by the most eminent philosophers of Europe.[14]

Certainly the lecturers in Paris included some very distinguished men. Heinzmann was convinced that in France were to be found all the leading exponents of chemistry, natural history, and mathematics and, with a few notable exceptions, he was right.[15] Under the Consulate it was particularly noticeable that many prominent men of science like Chaptal, Fourcroy, Laplace, Monge, and Berthollet were closely associated with the government.

There were various motives which prompted visitors to Paris to attend lectures on scientific subjects. Some sought enlightenment; for others it was no more than a fashionable pastime. A great attraction for some was the possibility of seeing a government minister such as Chaptal (Minister of the Interior, 1801–1804). Thus J. F. Reichardt, who traveled to Paris in 1802 in the hope of having his own opera performed there, was disappointed in his attendance at scientific lectures:

[14] H. R. Yorke, *Letters from France,* London, 1814, II, 22.
[15] Heinzmann, *Voyage d'un Allemand,* p. 149.

> It is annoying that scientists of reputation authorize the use of their name for the announcement of lectures which only take place very irregularly. On several occasions I went at the appointed times for the lectures of Fourcroy, Chaptal and other of the official princes of science, but all to no purpose. This irregularity may be a matter of indifference to those Parisians who have leisure and who are content to look from time to time at a Councillor of State or a Minister sitting in state in his professorial chair. For a stranger these unforeseen absences are vexing.[16]

Fortunately for Reichardt, Fourcroy gave several different courses of lectures and the German visitor was able to attend some at the junior military academy:

> The elocution of Fourcroy is more brilliant and animated. I have been present several times at his course of chemistry at the *Prytanée*. He arrived in his fine carriage and his imposing appearance formed rather a contrast with the dark and dirty room where he lectured in front of an audience of fine young men. . . .[17]

The Revolutionary period is of particular interest for the interplay of destructive and constructive elements. For science, a landmark in the phase of iconoclasm was the closing of the Académie des sciences on 8 August 1793. It may be clear to us now that the execution of Lavoisier on 8 May 1794 was in his capacity as a tax farmer and not as a man of science. That a quite different impression was given to many contemporaries is suggested by Yorke: "It was not to be expected that the murderers of Lavoisier would become the patrons of the arts and useful sciences."[18] When the French government became the patron of science, therefore, the situation struck visitors as something of a paradox, all the more so as science in their own

[16] Reichardt, *Un hiver à Paris*, p. 456.
[17] *Ibid.*, pp. 455–456.
[18] Yorke, *Letters from France*, II, 28.

country was seldom given government support. Bugge was most impressed not only by the existence of flourishing French scientific institutions but also by the fact that they were supported so generously from *government* sources. He interpreted this as "expiation" for the previous vandalism.

As most of the Danish astronomer's book is concerned with Paris and its institutions, some comment may be necessary on the author's original title, *Travels in the French Republic*. The original edition goes some slight way toward justifying this title by beginning with a detailed account of the author's journey from Copenhagen through northern Germany to Paris. This has been omitted in this edition on the grounds that the details are of little interest to American and British readers. In Chapter I of this book, therefore, Bugge has already arrived in Paris. There the effect of the work of the Convention and Directory had been to increase the centralization of the French state. In science, nearly all the important institutions and the leading practitioners were to be found after the Revolution in Paris. The financial support of science from the government was similarly weighted heavily in favor of Paris, as Bugge points out (p.159). In the foundation of the National Institute a deliberate attempt had been made to appoint, in addition to resident members in Paris, an equal number of correspondents in the provinces. The corresponding members, however, did not receive a salary and their status was very much lower than that of the resident (that is, full) members. Yorke's criticism of the centralized system was a more fundamental one:

> Every preference is manifestly given to the capital, even at the expense of the departments. 144 members [of the whole Institute] are always to be found in Paris, but the poor departments, containing a population of thirty to one, compared with the metropolis, are never able to produce more able men than the latter. This is absurd in the extreme; for everyone knows that under

> the old monarchy, there were men of the most distinguished acquirements scattered over the provinces, who were often equal and in many instances far superior to the members of the Parisian Academies.[19]

Bugge's information is not completely limited to the capital. His statements on the *écoles centrales* in the provinces, for example, is based on the direct testimony of the officials appointed to examine these schools.

The Danish astronomer arrived in Paris on 18 August 1798, but no formal duties were expected of him for at least another month. He was therefore able to make full use of this time to explore the French capital. Thus we find him on 28 August being introduced to a meeting of the First Class of the National Institute. On the last days of the republican year VI (mid-September 1798), he observed the first national exhibition of arts and manufactures and on the first day of the new year (22 September), he witnessed a balloon ascent at the Champ de Mars. But we have already delayed too long in giving some account of the author himself.

Thomas Bugge was in his late fifties when he was chosen by the Danish government to represent that country in the final stages of the establishment of the metric system. He was well qualified for his task because of his experience in astronomical and geodesic work. Born on 12 October 1740, Bugge had studied mathematics and theology at the University of Copenhagen. In 1761, as an assistant at the Copenhagen Observatory, he helped in observations of the transit of Venus. After some experience in geodesic work, he was appointed in 1765 chief land surveyor of the Kingdom of Denmark. His first book, an elementary work on mathematics, was published in 1773. In 1777 he succeeded to the chair of astronomy and mathematics at the University of Copenhagen and at the same time became Astronomer Royal at the Copenhagen Observatory.

[19] *Ibid.*, II, 38.

In 1777 Bugge published a work in Danish on surveying which was translated into German as *Beschreibung der Ausmessungs-Methode, welche bey den dänischen geographischen Karten angewendet worden* (Dresden, 1787). A significant contribution to observational astronomy was his *Observationes astronomicae annis 1781, 1782 & 1783. Institutae in Observatorio regio Havniensi et cum tabulis astronomicis compuratae* (Havniae, 1784). He also published a textbook of astronomy and one of mathematics: *De første Grunde til den sphäriske og theoretiske Astronomie, samt den mathematiske Geographie* (Copenhagen, 1796); *De første Grunde til den rene eller abstrakte Mathematik* (3 parts, Copenhagen, 1813, 1814). Both the latter works were also published in a German edition.

Denmark's greatest claim to fame in the history of astronomy is undoubtedly through Tycho Brahe (1546–1601). Bugge was the eighth astronomer after Tycho to be appointed to the Royal Observatory at Copenhagen. Another distinguished astronomer who had earlier held that post was Olaus Rømer (1644–1710). Rømer had been associated with the Académie des sciences in Paris in the 1670's, and while there had observed the satellites of Jupiter and had drawn the important conclusion from his observations that light travels with a finite velocity. Rømer then returned to Denmark and was astronomer royal for some thirty years. Minor pupils of Rømer succeeded him and the observatory went into a period of decline exacerbated by its destruction by the great fire of Copenhagen in 1728. On Bugge's appointment in 1777, he immediately began an extensive program of repair of the observatory, which was furnished with new instruments.

At the end of the eighteenth century, Copenhagen was the capital of the "twin kingdoms" of Denmark and Norway. Although it was a flourishing center of trade, it had been almost completely isolated from the effects of the French Revolution. Bugge himself had enjoyed the opportunity of travel abroad when in 1777 he had undertaken an extended tour of Germany, Holland, France, and England. His observations of France in 1798–1799, therefore, lack the naïveté of man's first impressions of life in a foreign country. His journey to France increased his reputation in Denmark and in 1800 he was elected secretary of the Danish Royal Academy of Science, a post which he held until his death in 1815 at the age of seventy-four.

In 1804 he was elected a corresponding member of the French Institute in the astronomy section. Bugge's official position undoubtedly gave more weight to a letter he addressed to Delambre that was read at a meeting of the First Class on 9 November 1807.[20] The letter described the recent bombardment of Copenhagen by the British fleet (2–4 September 1807), when some 2,000 persons were alleged to have been killed and the university quarter destroyed. Bugge's house, his library of 7,000 volumes, and his mathematical instruments were destroyed in the resulting fire. This letter aroused some general sympathy from the members of the First Class.

The Danish astronomer presents his observations on his visit to France in the form of "letters" and explains how these were compiled:

> I did not wish to publish my diary as it stood, for in it information on the same subject was necessarily to be found scattered in several places; therefore I gathered together from it everything that I had learnt about such subject at various times, so as to be able to present it to the reader under one heading and so, I flatter myself, the reader will miss nothing on the subject. The form in which these observations appear is really quite unimportant. I chose letter form because in a few letters from Paris to my friends, I had already described several things in detail. [21]

[20] Académie des sciences, Paris, dossier: Thomas Bugge.

[21] Author's Preface, pp. iv–v of the German edition. This preface is entirely omitted from the 1801 English edition.

For his journey to Paris, Bugge left Copenhagen on 29 July 1798 and traveled via Altona and Munster to Wesel. Crossing the Rhine, he entered the newly conquered territory of the French Republic. He passed through Louvain on his way to Brussels, where he spent two nights. He noted the scars of war at Valenciennes before continuing his journey south to Paris, which he reached on 18 August. It is here that we begin his account.

Chapter 1

EDUCATION AND SCIENCE

The Primary Schools and the Ecole Normale

By the law of 3 brumaire year IV (25 October 1795), the principle was established for the first time that education was a government responsibility. The many institutions for higher education described in this chapter therefore had extensive financial support from the state. Unfortunately in the sphere of primary education, where Bugge logically begins, the situation was very different.

Under the *ancien régime* education had been very largely the responsibility of religious orders. Although the expulsion of the Jesuits in 1762 had had a serious effect, it was secondary education rather than primary education which suffered. Other orders including the Oratorians and the Benedictines continued to play an important role in their schools. The abolition of the ecclesiastical tax in 1789 took away the financial support of church schools and this was followed immediately by the confiscation of church property and goods. By February 1791, primary and secondary education had almost ceased, although a few isolated priests continued private teaching. During the next three years a very large number of projects for the reform of education were discussed but little constructive was achieved until the law of 3 brumaire was passed by the Convention. By then some of the earlier democratic principles (for example, that the education of the poor and illiterate was a solemn duty of society) had been modified. Freedom was interpreted as the lack of compulsion to attend any school at all. Elementary teachers, where they could be found, received only lodging from the state. For their salaries they had to rely on tuition fees paid by the pupils. As classes were often small, state primary school teachers had little financial inducement for their work. Bugge shows that private instruction continued, although in a desultory fashion.

The poor primary education soon weakened the objectives of secondary education. It is surprising, however, that the superstructure of higher education was able to achieve so much with so little solid foundation. Of course the consequences of poor primary education take several years to affect the intake of a higher age group.

Although the Revolution produced an educational crisis, it also provided unparalleled opportunities. The gravest shortage was that of teachers, and it was particularly serious after the general neglect of education in the period 1790–1794. The many grandiose schemes for educational reform were useless without teachers and these had to be trained very quickly to meet the emergency. A new concept of "revolutionary" instruction had been introduced in 1794 to give training in Paris to men from all over France in the refining of saltpeter and the manufacturing of gunpowder and cannon. The course lasted ten days and was considered a great success. This provided a precedent for other programs of emergency instruction, of which the best known is the Ecole normale established by the Convention. Prospective teachers from all over France were to come to Paris for a four-month course during which they were to attend lectures by leading experts. Half the instruction was in scientific subjects. The students were then to return to centers in the provinces and on the strength of what they had learned in Paris form the staff of local teacher colleges. Unfortunately the high level of instruction presupposed that the students were already familiar with their respective subjects. Young Joseph Fourier, who was able to discuss higher mathematics with his professor, Monge, was very much an exception. The amphitheater of the Muséum d'histoire naturelle, although large, was only half the size required for the 1,300 to 1,400 students. That the students were completely heterogeneous in age, interest, and education is brought home to us by the inclusion of the veteran naval explorer Louis Antoine de Bougainville, then aged sixty-six! The Ecole normale in Paris opened its doors on 20 January 1795, and finally closed four months later on 15 May. An Ecole normale was reestablished under Napoleon by the decree of 17 March 1808.

The objectives of the earlier school and the way they were carried out were rather confused. The original objective had been to instruct in methods of teaching, but this idea was soon overlooked and the instruction given was in particular branches of knowledge. Two distinctive features of the Ecole normale were the use of the printed text of the lectures (obtained from stenographers) so that students could go over the lectures at their own pace. There were also "debates" in which the students could ask any questions about the course at special meetings with the professor. This attempt to democratize higher education had little success. Probably the most favorable verdict that could be given on the experiment of the Ecole normale was that it established an educational precedent of value for the planning of the University of France in 1806–1808.

In the account of Paris which I intend to give, you must not expect me to confine myself to chronological order; but I shall arrange in my journal all that I intend to say on the different subjects, under their proper heads—a method which, in some measure, will prevent repetition and disorder.

Primary Schools

I shall begin with public instruction. The first are called primary schools . . . where reading, writing, and arithmetic are taught. There are many private institutions erected in Paris, the object of which is to prepare youth for the higher classes, so that they may be transplanted from those schools to the central schools. These private institutions in Paris are in general conducted in a very proper manner; but I cannot say so much of those in the provincial towns, and in the country. Formerly the clergy claimed the exclusive right of instructing youth. The parish priests were allowed lands and houses, but being now deprived of these benefices, they are obliged, as their only means of support, to teach small schools, where the country people pay for the education of their children; but those schools are so little frequented, that the rising generation may be said to grow up without any instruction.

We may conclude, that the primary schools were very much neglected from the speech which Bitaubé, the president of the National Institute, delivered in the Council of Five Hundred, and the Council of Ancients, on the second complementary day of the Republican year VI (or 18 September 1798). I cannot in this place omit a passage in it, which reflects so much honour on the National Institute, and the orator who pronounced it. ". . . the central schools should not be deprived of their first and firmest foundation, the primary schools. I have already acknowledged, that this measure is very dear to your hearts. The Republic has cause to lament, that this important work has been suspended for a long time, from a series of unfortunate circumstances. We trust, therefore, to your wisdom, that you will fix their existence on a firm and immutable basis. The members of the Institute feel it their duty to declare the lively interest, which they take in every part of your deliberations and labours. The members of your Institute are deeply interested in the fate of these schools, and they are anxious that such measures may be adopted as will tend to multiply and fix them on a ground that will shortly evince the wisdom and utility of the measure.—But, citizen representatives, you know how important a thing it is for public order, the maintenance of the laws, and the correction and purity of morals, that those, whose fathers you are, should be early instructed and usefully employed. You are called on to watch over a race of young plants, which are now drooping—and, if not speedily revived, will fade away. The happy effects of the central schools are already experienced in different departments; the happy consequences of other public institutions are daily diffusing themselves. It is in your power to remove the misfortunes of which we complain; so that an active, aspiring, and ingenious people will have the pleasure of seeing their youth return once more to instruction, when it is held out to them."*

The presidents of both Councils in their answers, pronounced a panegyric on the primary schools . . . . Time will prove whether it would not redound more to the advantage of the French nation, that these patriotic views should be carried into execution, than the conquest of entire provinces. Without instruction, the rising generation will have to lament the fatal consequences of ignorance, immorality, and unbridled licentiousness.

Ecole normale

In consequence of the Revolution, everything was changed, and even the best institutions under the monarchy were subverted, or annulled, with the exception of the Collège de France in Paris, which has undergone no change. It was found necessary, that other institutions should be substituted in place of those that were abolished, and to which they gave the name of normal schools. In pursuance of the decree of 24 nivôse year III (15 January 1795) the National Convention ordained, that professors and teachers should be established, over all the Republic, and they gave the general name of Normal Schools to those schools, to which men of clear understanding only were to be appointed, to prepare youth for the higher schools. . . . In the first sitting, or assembly, the professors only spoke; in the subsequent ones, the subject was reserved, and all the pupils in succession were at liberty to deliver their opinions on it. They could put questions

***Comte rendu et presenté au corps législatif, le 2 jour complémentaire de l'an VI, par l'Institut National des sciences et arts*, Paris, an VII [1799], pp. 186–187.

to the professors, and the professors, in their turn, could question them; so that the subject of enquiry was generally sifted to the bottom; as there was no restraint on the freedom of discussion, except what good manners and politeness imposed.

The teachers were chosen from among men of the first talents, known either by their discoveries or writings. On the first and sixth day of each decade, Lagrange and Laplace taught mathematics, Haüy physics and Monge geometry. On the second and seventh days, Daubenton lectured on natural history, Berthollet on chemistry, and Thouin on agriculture. On the third and eighth days, Buache and Montelle read geography; history was given by Volney, and morality by Bernardin de Saint Pierre. The fourth and ninth days in each decade, were devoted to the principles of universal grammar by Sicard, logic by Garat, and general literature by La Harpe.[1]

The journal, which I have now before me, the National Convention ordered to be published. It consists of two grand divisions, lectures and debates, or conferences. Six octavo volumes of the lectures have already appeared.* These six volumes contain sixty-one collections and lectures of the professors just mentioned, in the main classes, from 20 January to 15 May 1795. In truth, whatever fell from the lips, or flowed from the pens of such enlightened men as Lagrange, Laplace, Haüy, Monge, Daubenton, Berthollet, Thouin, Buache, Volney,

[1]We may be duly impressed by the high standard of the faculty chosen for the Ecole normale. Lagrange and Laplace were possibly France's most able mathematicians. Haüy, one of the founders of the science of crystallography, made significant contributions to the teaching of physics in the Revolutionary and Napoleonic periods. Monge is usually regarded as the founder of projective geometry. He was able to give this a prominent place in the curriculum of the Ecole polytechnique. The veteran zoologist Daubenton had been a collaborator of Buffon at the Jardin du Roi. Berthollet had recently discovered the use of chlorine for bleaching and was beginning to work out a new concept of chemical affinity which threw a completely new light on the theory of chemical reactions. Thouin made important contributions to agriculture. The remaining names are almost equally distinguished in the humanities.

**Séances des Ecoles Normales, recueillies par sténographe et revues par les professeurs, Leçons,* tom. I–VI, Paris, an III [1795].

Sicard, and La Harpe, had a claim on the public attention; but they did not extend beyond the first principles of the sciences, which was as much as could be expected in four months, or twenty-four lectures of an hour each. In my opinion, Haüy has been very successful in his physical lectures. There is only one volume of the debates or conferences published;† it contains twenty-five collections; but it seems far from being interesting (perhaps it could not be otherwise) and it was very judiciously compressed into one volume.

The object which the Convention had in view, in erecting the normal schools, was to introduce and explain the methodistic mode[2] of instruction, as it is now called by some. On re-perusing the sixth volume of the works of the normal schools, I found nothing to complain of. It must strike the reader, however, that the normal schools can have produced nothing remarkable. They were raised upon a hasty and unstable foundation, and hence, in less than a year, they were dissolved.

The schools, which exist at present, are the central schools, the polytechnic school, and the schools for the public service (*Ecoles de service publique*).

Ecoles centrales

The radical innovations of the *écoles centrales* could only have been envisaged in a situation in which the previous educational system was in ruins. The education of the *ancien régime* had been intended to produce gentlemen, lawyers, and academics. The new education was intended to enable the citizen to understand the modern world. It was originally planned to include in the curriculum of these schools such subjects as agriculture, technology, hygiene, and modern languages, but even

†*Séances des Ecoles Normales. Débats,* tom. I. Paris, an III [1795].

[2]The original objective had been to instruct in *methods* of teaching rather than actual subjects (see editor's introduction, p. 21).

when this project was abandoned the curriculum was revolutionary enough.

According to the then-dominant school of philosophy in France, that of the *idéologues*, liberty is a natural state and a free man requires education. His ignorance can lead to error that might affect the well-being of the society in which he lives. Following Condillac, the *idéologues* believed that language is one of the keystones of education, for all we know is through sensations and these are expressed by words.

There were several reasons why science was given a place of central importance in the curriculum of the *écoles centrales*. Observation and principles of nomenclature were basic considerations for the *idéologues* and both could be cultivated in the study of natural history. Natural history was also valuable as a means of acquiring an understanding of nature. It was hoped that the study of physical sciences would eliminate superstition, for it was argued that a person who understood the laws of physics could not believe in miracles. The teaching of mathematics, physics, and chemistry was seen as a means of creating habits of precise methodical thought.

Unfortunately the *écoles centrales* did not provide an integrated curriculum but rather a series of isolated courses. It should not be assumed that pupils followed from one group of subjects to the next. Most pupils only stayed one year and followed a single course—drawing and mathematics being particularly popular. In general their curriculum supposed a higher level of attainment than was provided in the primary schools. Even the reading, writing, and arithmetic which were supposed to be taught in the primary schools could not be taken for granted.

Although the idea of *écoles centrales* came from Paris, as did the broad basis of the curriculum, the actual running of the schools was in local hands. There were therefore significant differences between *écoles centrales* in different departments. It is of particular value to have a more detailed account of a particular *école centrale*, one of the three actually established in Paris. Needless to say, the standard here was higher than in the provinces. It was particularly fortunate, with its staff including, for example, the mathematician Lacroix and also Fontanes, who was to become Grand Master of the University in 1808. An example of the high level of science taught at the Ecole centrale des quatre nations is provided by the knowledge expected of pupils graduating in the physics class.[3] This included not only the comparatively straightforward knowledge of statics, dynamics, hydrostatics, and optics, but also more complex problems in hydrodynamics. Pupils learned about the precession of the equinoxes and also the theoretical basis of the metric system.

Generally the *écoles centrales* constitute an important landmark in the history of education. Yet in practice they did not flourish and in 1802 they were abolished. The reasons for their failure are complex, but a few causes may be listed here. Basic perhaps was the fact that the French people did not accept the philosophy of the *idéologues* on which the schools were founded. The curriculum implied a fair standard of preparation, but many of the pupils were hardly literate. There was an abysmal shortage of suitably qualified teachers and many posts were not filled at all or were filled by poorly qualified teachers. The administration was poor and the financial provision for the teaching of practical science was usually quite inadequate.

In conclusion, we may quote a judgment on the *écoles centrales* by a German idealist, G. T. von Faber, who went to France after the Revolution to offer his services and obtained a post in the civil administration:

> Lakanal and Chenier explained the motives of the law

[3] *Exercise des élèves de l'Ecole Centrale des Quatre Nations . . . sur la physique, la chimie et les applications des mathématiques à la physique . . .*, Paris, an X [1802].

of 3 brumaire year IV which established two degrees of instruction by means of *primary* and *central* schools. In point of fact, however, the former never existed from the want of means to pay the master; and the latter languished, because, through the failure of the others, the elementary branches of instruction were deficient. In some departments the central schools had more professors than pupils; in all, without exception, the number of students fell far short of that which the population should have furnished; and if some of them were tolerably well attended, it was only because the teachers, in order to enlarge the sphere of their utility, made up for the deficiency of the inferior seminaries and descended to the drudgery of elementary instruction. The academic method pursued in the central schools, after the model of the German universities, might have sufficed for youth of riper years and greater proficiency, but was by no means adapted to childhood, which requires incessant superintendence and an unremitted course of discipline. If elementary instruction was neglected, at least the various branches of mathematics, natural philosophy and natural history became favourite studies, through the exertions of some of the central schools. The want of a more complete system of instruction was meanwhile most sensibly felt, and gave rise to a great number of private institutions, many of which, in Paris and likewise in the departments, are eminently distinguished; but these are accessible only to wealth, and the great bulk of the people, to whom fortune has deprived this advantage, remained without instruction.[4]

The law for the central schools was enacted on 3 brumaire year IV. The regulations are as follow: There shall be a central

[4]G. T. von Faber, *Sketches of the Internal State of France*, 2nd ed., London, 1813, p. 141.

school in each department. The whole of the instruction shall be divided into three parts or sections; drawing, natural history, the ancient and modern languages, shall be taught in the first; mathematics, physics, and chemistry, in the second; and universal grammar, *belles lettres,* history, and legislation in the third. The pupils to be received into the first at the age of twelve, into the second at fourteen, and into the third at sixteen. There shall be a public library in each central school, with a botanic garden, and apparatus of chemical and philosophical instruments. The professors to be examined and chosen by a Jury of Instruction (*Jury d'Instruction*) and the choice to be confirmed by the departmental administration. A professor cannot be dismissed by the aforesaid administration, unless there be a complaint preferred against him by the Jury of Instruction, which must be well grounded, as he is at liberty to defend himself, and there is a final appeal to the Directory. The salary of the professor is from 2,400 to 3,600 francs, also to be paid by the departmental administration. They have besides, such a yearly gratuity from each pupil, as the department thinks fit, which seldom exceeds twenty-five francs. The fourth part of the pupils are in general too poor to spare anything.

It is easy to remark, that the general rules or laws are very well digested; but the manner in which they are to be obeyed or maintained, should have been laid down at the same time. It is to be lamented also, that morality is passed over; especially as the public exercise of religion is abolished. In the second section, the learner from fourteen to sixteen, is instructed in the abstract sciences, which tend very much to sharpen the understanding, and to call forth the latent powers of the mind; and from sixteen to eighteen, he is taught to read the best historians, a study peculiarly improving to the minds of youth at that period.

From the central schools I shall now proceed to the *Ecole centrale des quatre nations,* established in the buildings of the former *Collège des quatre nations.* I shall give you an account of the teachers and the hours of lectures. (Lectures are given every day, except the fifth and tenth days in the decade.)

First Section: Ancient Languages

Gueroult, the elder, reads from nine to half past ten in the morning.

Natural History

Brongniart the younger, from half past ten till noon. He is a lively young man, has a pleasing delivery, and I have listened to him with a great deal of pleasure.

All the pupils in this class have the afternoon to themselves; and it is entirely at their own option, to repeat or not, what they heard in the morning.

Drawing

Moreau, the younger, teaches drawing from twelve till half past one.

Second Section: Mathematics

Lacroix teaches arithmetic, algebra, geometry and trigonometry, from nine to eleven in the morning, on all the odd days (the first, third, seventh and ninth days of the decade). Lacroix has a fine delivery, and is a very good mathematician, as is well known by [his books on trigonometry, descriptive geometry and calculus].

Experimental Philosophy and Chemistry

Brisson reads all the even days, from half past ten to eleven. He is an impressive reader, and all his reasonings are well grounded. He is known by a work on the specific gravity of bodies. He has besides, written three volumes on physics, two of which are already published, and the third is impatiently looked for. I do not hesitate to say, that this work contains the best system of physics in the French language. In this section there are only two hours each day set apart for reading lectures; so that the pupils have time enough to learn mathematics and physics in the second year, if they choose to occupy their time in such studies.

Third Section: General Grammar and Logic

Domergue reads all the odd days, from nine to eleven.

History

Mentelle, all the even days, from nine to eleven.

Legislation

Grivel, all the even days, from nine to eleven.[5]

Belles lettres

Fontanes, all the even days, from eleven to one.

This school has besides an agent and secretary, Lepine.

This school has a handsome library, which formerly belonged to the *Collège des quatre nations*; a collection of philosophical instruments, which are rather old, but kept in good order by Brisson, on which he makes experiments very successfully. There is likewise a small botanic garden annexed to it. This school, when a college, was mouldering fast into ruins; but it is now undergoing a thorough repair, and, when finished, will be found to be very neat and convenient.

The second central school in Paris is in the Panthéon, formerly the church of Sainte Geneviève. The regulations are entirely the same. Among the teachers in natural history are Cuvier, and Deparcieux, who is still better known. The third central school is in the suburb of Saint Anthony, in the former Jesuits' College. Among the teachers in those schools some are known by their literary productions, and those who are not, may yet be very well qualified to fill their respective situations. These two schools have good libraries, a collection of at least the most useful philosophical instruments, and each a small botanic garden.

In those departments where universities, colleges, large cloisters, palaces of emigrants, and libraries were already established, it was easy to organize central schools; but where such universities, &c. were wanting, they are not even at this day furnished with central schools. . . .

Besides the three central schools in Paris, ninety-seven are intended for the departments, of which fifty-one are organized, and forty are yet unorganized. Different teachers are still wanting in some of the organized schools; for example, at St. Girons, in mathematics and physics. In Tulle, all the teachers are wanting, except those of drawing and grammar. In Monté limar, there are

[5]According to this timetable, students would not be able to attend all the courses. This was also the case at the *École centrale du Panthéon* (*Almanach National pour l'an VII*, p. 503).

only two, one in natural history, and the other in mathematics. In Châteauroux, there are none in physics, or in the whole third section. In Puy, there are none at all in the first class. In Porentruy, Anvers, Nevers, Pau, Autun, physics and chemistry are quite neglected, for want of professors. Collections of instruments and libraries are wanting in many. There are no teachers of the foreign languages to be found in any. Lalande, since his journey to Gotha, last summer, confesses that the knowledge of German literature would amply repay the trouble of acquiring the language of that country, even to be able to read the books which appear in it. He has written to the Minister of the Interior on that subject, and entreated that persons skilled in the German language, may be appointed to teach it in the central schools. Yet we do not observe the least inclination to learn the foreign languages. In the narrow circle of my acquaintance, however, I know some who speak German with fluency. Among this number are Mr. Bourgoing, well known by his justly admired writings on Spain; Cuvier, member of the National Institute, and professor of natural history; Coquebert, professor of history, and a member of the general department of weights and measures, a young man of very genteel address, and good education. He recommended the introduction of many articles of utility; but, in pursuance of the Minister's advice, he went to Italy, where he exchanged his pen for a sword, and is now a good soldier.

I have already remarked, that morality and geography are not ordered to be taught in the central schools. One teacher only is appointed for Latin and Greek, to which he devotes two hours each day, the age of the pupils being from twelve to fourteen. But in so short a time, pupils of that age cannot be expected to make any great progress in the acquisition of those languages. I have heard many of the best philologists in Paris complain that ancient literature is very little attended to, not to say quite neglected. In some countries, it is prized beyond its value, and in others, it is depressed beneath it. In my opinion, the lovers of science ought to know at least as much Latin and Greek, as will enable them to trace the roots of those scientific words, for which we are indebted to those languages. Lectures are read in the central schools; but no books are prescribed to the pupils,[6] nor are they called upon to repeat what they have heard. I am not quite certain that a youth, from the age of fourteen to sixteen, can be well grounded in the principles of science, by this mode of instruction. I have been present at the public examinations, and found that most of them knew some things in a general way; but that very few were masters of the primary principles of science.

Towards the close of the republican year, the Directory appointed commissioners to travel through the departments, in order to examine, and to make a report of the state of the central schools. Many of these were my friends and acquaintances, and they assured me, that in most places, they found those schools in a very indifferent state: even some of the teachers knew very little of what they professed. The commissioners saw that it was very necessary that proper books should be written for the use of these schools. They lamented, at the same time, that, in most of the departments, the central schools were little sought and attended by very few.

As soon, however, as defects can be supplied, and proper regulations adopted, with the means of carrying them into execution, it is very probable that those central schools, such as they are, will be found to be of great utility.

Ecole polytechnique

The Ecole polytechnique was founded as the Ecole centrale des travaux publics on 24 September 1794. On 1 September 1795 the name was changed to Ecole polytechnique. Although there had been both military and engineering schools under the *ancien régime*, the principle on

[6] Bugge's generalization that "no books are prescribed to the pupils" is rather too sweeping. As early as 13 June 1793 a project had been advanced for the composition of elementary books. Although no action was taken immediately, by the time of Bugge's visit repeated efforts to commission elementary works suitable for the *écoles centrales* were beginning to bear fruit.

which this new institution was founded was that civil and military engineers could be given a common basic training. Nevertheless the curriculum was never completely vocational and it was hoped that the training provided would create "enlightened citizens." It should be noted that entrance was by competitive examination and the high standard of teaching became widely recognized. The curriculum described here placed emphasis on *practical* training in science. Its utilitarian aspect was modified in the reorganization of 1799, which placed greater emphasis on pure science in the curriculum. For the first ten years the school was situated at the Palais Bourbon, but in 1805 it was moved to its present sight on the Montagne Sainte–Geneviève. The most important change in the early history of the Ecole was effected by the decree of 16 July 1804 which put it under a military regime. The uniforms worn by students at the school are often associated with the period of militarization, but the account given here reminds us that uniforms were introduced before this. Although Deshautchamps was director of the school when Bugge visited it, more famous and influential as directors were Monge (1797–1798, 1799–1800) and Guyton de Morveau (1798, 1800–1804). There is extensive literature on the Ecole polytechnique, but the following provides a valuable eyewitness account of the practical facilities at the school as well as the functioning of lectures and examinations.

The very practical character of the second year course requires the comment that this had been one of the justifications for founding the school under the name Ecole centrale des travaux publiques in 1794. A notable feature of the reorganization of 1799 was the reduction in the number of practical courses. Raymond-Latour, employed at the Ecole polytechnique in the period 1795–1798 as a demonstrator in chemistry, gives an interesting firsthand account of the generous facilities which had been made available by the Convention:

My admission to the Ecole polytechnique brought a great improvement to my situation and temporarily put an end to my anxiety about the future. No educational institution up to that time had offered such a remarkable example of magnificence and liberality; nothing had been spared to make this school celebrated. One might have said that the government of that time had the simple preoccupation of filling the great gap which several cruel unsettled years had brought into education and making people forget completely about them. The professors and their assistants were comfortably lodged with good heating and they were given generous salaries. The students, abundantly provided with all the material necessities for their work, could devote themselves more successfully to study. . . . Only the higher sciences were taught in this school and the professors were men chosen from among the elite of scientists. For chemistry there were three large laboratories for the lessons of the [three] professors with a demonstrator attached to each of them as well as a general demonstrator responsible for chemical supplies. In addition, there was a laboratory for each division of the students. They repeated there under the direction and guidance of the demonstrator the principal reactions which had been the subject of the lecture. . . . The chemicals and the apparatus necessary for their experiments were given to them on presentation of a voucher signed by me. Seeing so much prodigality in the support of chemistry one was tempted to ask oneself if the reign of the adepts was about to begin again and if the philosophers' stone might finally appear from so many furnaces.[7]

Raymond-Latour was not the only one to be impressed

[7] J. M. Raymond-Latour, *Souvenirs d'un Oisif*, Lyons & Paris, 1836, II, 47–49.

by the extensive facilities available at the Ecole polytechnique. The lawyer Friedrich Meyer from Hamburg was concerned at the lavish expenditure on the new foundation which, he thought, exposed it all the more easily to the attack of the critics of any such elitist institution:

> The great luxury (one might say the excentric luxury) with which the Directory with a special and very just predilection supports this new foundation seems to me to constitute a danger to the permanence of this institution. Complaints have already been made on several sides against excessive and useless expenditure and the government has already begun to cut out everything that is not essential. In this category we would place the 24 laboratories intended for the individual work of students in which large sums went up uselessly in smoke before these young men had acquired a degree of knowledge great enough to extract any great benefit from these costly experiments. These laboratories have been reduced to eight, which already produces a considerable saving.[8]

The next school, but of a higher order, is the Polytechnic School, in the former *Palais de Bourbon,* where the Assembly of Five Hundred also hold their sittings in a large hall. The pupils are translated from the central schools, after a preliminary examination, in the elements of arithmetic, algebra, geometry, trigonometry, &c. The number of pupils is settled at 360, who are divided into brigades, twenty to each hall, under the inspection of the teachers, and a visitor, or chief inspector, whom they alternately choose from among themselves. The common courses in these schools require three years, and the school is divided into three corresponding classes. Deshautschamps, the present director of the Polytechnic School, is a profound mathematician. He is often present at the lectures, and spares no pains to keep the

[8] F. J. L. Meyer, *Fragments sur Paris,* trans., Hamburg, 1798, II, 87.

pupils in proper order. The last year, he carried a decree that the teachers and pupils should wear a uniform of buff-coloured waistcoats, and blue frocks with yellow buttons, on which are inscribed the words, "Ecole polytechnique." Besides the director, there are two administrators, Le Brun and Lermina, who are very honest men.

First Year, or First Class.

In this class, the higher algebra and analytic geometry are taught, together with that part of geometry which is particularly applicable to the practice of stone-cutting, carpenter's work, sciagraphy, or shadowing, perspective, and the construction of maps. The teachers are Monge, who is now in Egypt, and his assistant Hachette. The chemistry of Fourcroy is also explained in his class, by Salternes. Hassenfratz lectures on general physics, including mechanics, and the other parts of physics, which are found necessary in mechanical employments.

Second Year, or Second Class.

The arts of laying out roads, erecting bridges . . . are taught by Lagarde and Dubois. The science of building, or any peculiar style of architecture is taught by Gay de Vernon. Prony and Fourier explain hydrostatics, hydraulics, and mechanics. The chemistry of the organic, vegetable, and animal substances, are taught by Berthollet and Chaussier. The former is at present in Egypt.

Third Year, or Third Class.

Fortification is taught by two officers of the engineers. The construction of such engines as relate to mechanics, are explained by Prony and Fourier. The chemistry of minerals, metallurgy, and mining, are taught by Guyton de Morveau. Lagrange besides, reads lectures on differents parts of the mathematics, particularly the analytic. There are three drawing masters.

The Polytechnic School is kept in very proper order; it contains good philosophical apparatus in three rooms on the third story. In the first room, are three ovens or stoves, with glasses, to make experiments on the nutrition of plants by gases, and many conveniences for the prosecution of physical and chemical researches.

In the second room there is a large collection of mechanical and hydrostatical, optical, astronomical, electrical, and magnetical instruments; most of which belonged to Nollet, and Sigaud de la Fond; and they are kept in good order and well arranged.

Among the few English instruments, I observed there, was an excellent air pump on Smeaton's plan, improved by Nairne. This formerly belonged to Lavoisier; but, as it had only one tube, he exchanged it for one with two, which, though more quick, does not evacuate the same quantity of air in a given time. There is on the second floor, a hall highly decorated, which is filled with a great number of instruments and models, many of which ought to be in the first. A person is appointed to keep those instruments in order, and to arrange the new ones. The pupils have access to them when they please.

This Polytechnic School has a very neat and good library of about ten thousand volumes, of the chief works on the different subjects taught in this institution. It is open, for the use of the pupils, some hours every day, and on the decades the whole day. It is constantly consulted by the students, of whom I have often found from twenty to thirty in it at a time.

In a room, set apart for that purpose solely, there are models of machines, some of which are very interesting and useful; but others are of little value, and indifferently executed. It may be only called the beginning of a collection of that kind, which will be supplied by degrees, with models of machines of more importance, and better workmanship.

All these models, machines, and philosophical instruments, may be said to have cost nothing, having been partly taken from the former public collections, and partly from the royal philosophical and mechanical cabinets, or from those of the emigrants.

Three rooms are set apart for architecture alone. In the first, was stereotomy, which, in the scientific language of the Polytechnic School, signifies that part of stone-cutting, on which Frezier and De la Rue have written so much. The theory and rules of projection are first studied; as when a solid body, of a given figure, is to be cut, according to plans or schemes of a given position, such as a cylinder to be cut by another cylinder, or by two cylinders; or when a body, of which one end is a circle, and the other an ellipse, is to be cut by a given plane, there to define the curve lines of the projection; or, on a kind of cone, the basis of which is an ellipse, to define the section, which will be a circle. These, and many other such problems, are executed from models remarkably good.

This stereotomy, together with descriptive geometry, are cultivated with a great deal of zeal and industry. I will not say that the pupils should be ignorant of these things, nor will I deny that the knowledge of them may be found useful in many respects, particularly in the construction of charts and maps, designs in architecture, and mechanics, &c.; but I think I may venture to assert that they cost more time and application than they deserve.

The second architectural room was entirely appropriated to stone-cutting, or the determinative formation of certain figures. They make use of a composition of stone to form models of portals, gates, bridges, &c. which the pupils must work themselves. Most of those models were very neat, and on the whole well executed. Their height was from eight to sixteen inches.

The third room contained models of all the orders of architecture; and of entire façades, buildings, palaces, and temples. There is a person in the school, who models with great exactness and elegance.

The drawing or designing school is a fine long room, to which the light enters from above. It is divided into three classes. The first is confined to the drawing of heads, hands, feet, &c.; in the second, whole figures are drawn after designs; and in the last from the life, and from fine models in gypsum, of which the school has a remarkably good collection. Some very fine designs of the pupils are hung up in both.

The Polytechnic School has two very large and fine chemical laboratories, besides two of inferior extent, and some mechanical workshops. The director and administrator have lodgings, at free cost, in the school.

As a stranger, I have attended several lectures, among which were those on analysis by Lagrange. Whatever this great man says, deserves the highest degree of consideration; but he is too abstract for youth. In the examination of these lectures, it has been found, that he has discovered a new demonstration of the first principles of the differential calculus; and his *Solution des équations numériques de tous les degrés* (Paris, 1797) merits attention.

I have heard Prony's hydraulic lectures, particularly on the motion of fluids through pipes, and on the undulation of water. This extraordinary man has the most impressive and captivating delivery, which can possibly be conceived. In the course of the last year, he printed a text book of his lectures, containing theorems and problems, relating to his subjects, and a sketch or skeleton of the lectures themselves. In the year VII, Prony began a course, in which he proposed to demonstrate hydraulic theories in general. I have heard some of those lectures, which were excellent; but I fear that few of his hearers (about twenty in all) will be able to keep pace with him.

I have heard Fourcroy read on the fermentation of wines, and on the nature, quality, and preparation of alcohol. He made different experiments, to shew that the flame of burning alcohol contains a variety of colours, such as purple, violet, and green; the last of which appeared on mixing with it a solution of vitriol of copper. Fourcroy's delivery is fine, orderly, and emphatic; but perhaps a little too rapid for some youths beginning the study. When he had finished, he proceeded, in pursuance of a certain order, to examine from eight to sixteen pupils.

I have heard Hassenfratz lecture on electricity, lightning, and thunder. He concluded with an historical detail of all the systems of electricity; but passed over Symmer's theory,[9] or the dualistic system entirely. He adopted the theory of Aepinus, which has become prevalent in Paris. Haüy has since attacked the system of Aepinus, relative to electricity and magnetism. He

[9] In 1733 Charles Dufay stated that there are two kinds of electricity: *vitreous* and *resinous*. The first could be produced by rubbing glass, the second by rubbing amber. This dualistic theory was challenged by Benjamin Franklin in a famous letter of 1752. Franklin considered that there was only one kind of (static) electricity, which existed in two states, which he called *positive* and *negative*. The theories of both Dufay and Franklin were challenged in 1759 by Robert Symmer, who put forward a compromise theory. According to him there are two electric fluids,

[Hassenfratz] denies that the peals or claps of thunder proceed from the electric spark, which flies from one cloud to another, and bursts or strikes through the interjacent air, and insists that it comes from a vacuum, produced by the condensation of exhalations, which are converted into rain: If so, there never would be any peals or claps of thunder, which would not be accompanied with rain. I have also heard Hassenfratz lecture on machines.

The object of those lectures ought to be whatever relates to machinery, practical mechanics, and the different modes by which the motions of the machines can be made to produce the different effects, so as to attain the object. I have not heard enough of those lectures to enable me to say how far this object may be attained. Hassenfratz is deficient in delivery. Once in each decade, he conducts his pupils to see the machines, the management of the manufacturers, the rooms where the arts are carried on, and where mechanics work. I accompanied him, in one of his mechanical excursions, which are exceedingly useful, and furnish the pupils with ideas, which they could not obtain in lecture halls or libraries.

It was peculiarly enacted, that each of the pupils should have 1,200 livres a year, but this was decreed in the times of the *assignats*: so that those 1,200 livres in paper yielded very little money, and notwithstanding the *assignats* are called in, the pupils received little more than 200 livres a year, which amounted for the whole school to 72,000 livres a year. The Minister of the Interior, in the year VII desired the sum of 394,133 francs, for the use of the Polytechnic School; and certainly the pains and expense of the Government are well bestowed on an institution, which will furnish the state with so many public servants, and useful subjects.

essentially different from each other, and an electric charge consists of the accumulation of an excess of one of these fluids.

The theory of Franz Aepinus (1759) was a development of Franklin's theory of positive and negative electricity. A commentary on Aepinus' views on magnetism and electricity was published by the Abbé Haüy in 1787. A major contribution by Haüy to electricity was his study of the production of electricity in crystals by friction.

Hassenfratz was one of the less distinguished members of the faculty at the Ecole polytechnique.

When the lectures are closed, which happened this year in brumaire, there is an examination of all the pupils who have finished their course, and who would wish to enter into the schools destined for the training of candidates for the public service, in the construction of roads and bridges, ship-building, &c. or of those who would wish to become masters in other useful arts. For the present examination, the Directory appointed Laplace and Bossut. The first examined the students in analytic sciences, and the other in mechanics. Those who were to be examined were called up in order, and were obliged to demonstrate without book the proposed theorems, and to solve the problems on a black board; which was considered at once as a proof of talents and readiness. Laplace proposed questions in series, logarithms, and curved lines, in that part of algebra which is applicable to geometry and trigonometry, and in the differential and integral calculus. He proposed every question with much perspicuity and precision, and gently recalled the pupil to the right point, if he happened to wander from it.

Bossut, in another room, examined in mechanics, statics, hydrostatics, hydraulics, &c. I found most of the pupils answer very well, and with great readiness, difficult problems of the higher mathematics. But it must not be expected, that amongst so many, some would not be found of moderate and some of indifferent talents. Deshautschamps, the director, told me that Laplace, on the whole, was not well satisfied, and that some of the pupils were not entitled to that attestation, by which alone they could be admitted into the Schools for the Public Service. He lamented, and not without reason, that in those examinations, the young men were left without any occasional assistance to their memory or conception, especially when they found themselves bewildered in algebraic calculations. It is certain that a wink would often set them right, provided they had understanding and knowledge enough to avail themselves of it, which in itself would be a proof that they had not mis-spent their time. I informed Deshautchamps, that with us public examinations were held in gunnery, navigation, land-surveying, &c., that part of these examinations were by word of mouth, and part in writing, that all the abstruse theorems and problems were proposed in writing, to which the candidate was required to give written answers, and that this method allowed him time to reflect on the subject, to arrange it in his mind, and to revise and correct his piece as often as he pleased. Deshautschamps highly approved of this mode, and said he would spare no pains to have it introduced. These examinations were public, though I very seldom found that foreigners, and those who were not in some measure connected with the Polytechnic School, were present.

Schools for Public Services

Certain higher technical schools were founded or reestablished as a separate part of the work of the Convention's Committee on Public Instruction. If we exclude the Ecole polytechnique, which was not really a vocational training center, the remaining schools (military, naval, mining, civil engineering) were mainly creations of the *ancien régime*. There was now an opportunity to rationalize them and introduce some uniformity of entry by competitive examination. The Ecole polytechnique, which had originally been envisaged as an equal partner of, for instance, the Ecole des ponts et chaussées, soon became the normal means of entry to it. This had the effect of making the curriculum of the Ecole polytechnique a more general one. Details of the application of sciences taught there to particular problems could be left to the higher technical schools (*écoles d'application*).

The history of the Ecole des ponts et chaussées goes back to 1747, although it was not given this name until 1756. A formal constitution for the school was drawn up by Turgot in 1775. Under the *ancien régime*, however, instruction was not of a high standard. It was only after the Revolution that there was a regular teaching staff and the principle of entry by competitive examination was introduced. The reputation of the school soon became so high that many of the best graduates of the Ecole polytechnique went on to the Ecole des ponts et chaussées. Students were trained in the theory and practice of the construction of highways, bridges, and canals, and in quantity surveying.

The foundation of a mining academy at Freiberg in 1765 had given the French government an incentive to establish a mining school in France. In 1778 Necker approved the creation of a chair of mineralogy applied to mining to be established at the Mint. This became the basis of the Ecole des mines, which was formally constituted in 1783. By the time of the Revolution, however, it was already in decline. In 1794 it was reestablished in the rue de l'Université. Yet it was felt that the training given was too theoretical and in 1802, when Chaptal was Minister of the Interior, the Ecole in Paris was transferred to mining districts of the Saar and Savoy. The advantages gained by the students of early practical experience was offset by the difficulty of obtaining high quality teaching staff in the remote provinces.

The Geographers' School, like that of mining, had an obvious commercial and economic value. The value of maps in peacetime administration was enhanced by the requirements of war. Under the *ancien régime* training in map-making had been included in the instruction given at the military school at Mézières. Cassini de Thury had also trained a few men. The creation of this school by the Convention marked the recognition of a formal centralized scheme of training. Yorke commented,

> In this school the science of Geography is well taught, theoretically and experimentally and the mode of drawing maps and plans scientifically explained. Twenty pupils are admitted upon this establishment, with salaries, after they have completed their studies at the Polytechnic school.[10]

Military and naval academies had also been in existence before the Revolution and had produced some notable men, not only in the military field (like Wellington and

[10] Yorke, *Letters from France*, II, 16.

Napoleon Bonaparte) but also men who figure prominently in the annals of science, notably Laplace and Fourier. Visiting the school for military engineers, one English commentator was particularly impressed by the system of passing out examinations, and what he says applies also to the career structure of some of the other *écoles de services publics*:

> The pupils undergo a strict examination respecting the objects of instruction. . . . This examination is entrusted to a *jury* (as the French term it) composed of the commander-in-chief of the school, a general or field officer of the corps, appointed every year by the Minister of War, and one of the permanent examiners of the Polytechnic School. This jury forms a list of merit, which regulates the order of promotion. Can we then wonder that the French have the first military engineers in Europe?[11]

In the preceding letter, I have given an account of the Central Schools, and of the excellent management of that called the Polytechnic. When any one has punctually attended the Polytechnic School for a term not less than one year, and undergone an examination, he is then admitted to some of the "Schools for Public Services" or, as they are sometimes called, "Schools of Application." The pupils, soon after their appointment, obtain a small emolument, and afterwards pass from thence to the service of the state, when an opportunity occurs. These Schools of Application are for the construction of roads and bridges, for mineralogy, geography, ship-building, artillery, fortification, and nautical affairs.

Ecole des ponts et chaussées

The School for constructing Roads and Bridges, is situated in the Rue Grenelle, and was formerly a palace of the Duke de la

[11][Blagdon], *Paris as it was*, II, 297.

Chatel. It is disposed and embellished with consummate taste and magnificence; and contains a number of excellent apartments: the style, indeed, of the building, especially of the two rooms in front, is not very consistent with the modesty of a public school, but this congruity would be dearly purchased, by reducing the grandeur of the edifice to a level with its present use. Trudaine was, in the time of monarchy, the first founder of this school. Peronnet, author of an excellent work, entitled *Description des Projets et de la Construction des Ponts,* has since greatly contributed to its improvement. The busts of both those able men, the first of bronzed plaster, and the other of marble, are set up in the school-room.

Two of the apartments are appropriated to the museum, in which are not only drawings, but also models of buildings and machines, which relate in any respect, to the construction of roads and bridges, such as all sorts of rammers for driving vertical and inclined piles, five different models for sawing piles under water, in imitation of an English machine, which is very simple, cheap, and certain in its effects; various models of machines for raising water, of forcing pumps, and of sluices for canals; also models of the most remarkable bridges on the large rivers of Europe, of bridges formerly built in France, chiefly by Peronnet, and of the Pont Neuf at Paris, which is built very flat, and is uncommonly strong; models of the bridges at Neuilly, Nantes, Orleans, Brunoy, Nonnettes, Bicheret, &c. together with draughts and models of everything that relates to nautical architecture, some of the most remarkable of which are the caissons at Cherbourg. These have not fulfilled the public expectation; because, as Prony supposes, they have been badly executed. Though however, the basins are small, and completely filled, yet the anchorage within the cones is tolerably secure. All these models of the school for the constructing of roads and bridges are neat, accurate, and excellent: they are placed in the most beautiful order, and there is no country whatever, with a finer and more complete collection.

The school has a fine library of about two thousand five hundred volumes of good mathematical treatises, chiefly relating to hydrostatics, hydraulics, water-works, roads, and bridges. In the four rooms for instruction, the students are taught the elements of physics and mathematics; and to draw plans and sketches of roads, bridges, canals, harbours, and all kinds of buildings connected with them. They also learn to superintend the actual construction of buildings, to manage the expenses, and take an account of the annual rents. The number of students was fifty, thirty-six of whom had a pension of seventy francs per month, or eight hundred and forty francs per annum. Their course of study is usually completed in two years; before they leave the school, and frequently while they reside there, they undergo examinations, and are obliged to resolve problems and questions, relative to the practical part of their profession. Prony showed me some of these questions; most of which were difficult, and related to roads, bridges, sluices, &c. The solutions were accompanied by excellent drawings of plans and sections, and with exact calculations and circumstantial accounts of the expenses attending them. The present managers are the directors Chézy and Prony, and the inspector Le Sage, who exert themselves to the utmost, in preserving everything in a state of order and activity.

Ecole des Mines

The Mineralogical School, No. 293, Rue de l'Université has a large and rich collection of minerals disposed in glass[12] cases. The collection occupies six different apartments, and is divided into the three following classes: 1. The docimastic collection. 2. A geographic collection of French minerals. 3. An oryctognostic or systematic collection, illustrated by models in wood of the principal varieties of crystallization, after the system and discoveries of Haüy. The geognostic collection is included in the three former. There are also deposited here collections of drawings and models of mines, and of tools, instruments, and machines for the use of miners. The appointed number of students is twenty. Haüy is keeper of the cabinet of minerals, and

[12]"glazed"—J. J.

Vauquelin superintends the chemical department. Clouet is librarian and teacher of the German language.

The lectures for the year VII, or winter of 1798–1799 are the following:

Baillet, inspector of the schools, lectures every first and sixth day of the decade, on the art of mining. Hassenfratz, inspector, lectures every second and eighth day on mineralogy and metallurgy. Vauquelin, inspector, lectures every third and seventh day on docimacy. Brongniart, the younger, lectures every second and eighth day on mineralogy. Cloquet teaches drawing, Clouet the German language, and Leroy descriptive geometry.

The celebrated and learned Dolomieu is both inspector and lecturer of the mineral school; but he and two of the best students are in Egypt. The lecturers deliver their instructions in winter only; the summer being generally spent in making experiments at the laboratory, or in making tours to the various mining districts. This school has an admirable laboratory belonging to it, under the inspection of Vauquelin; the same who, in conjunction with Fourcroy, has made so many noble discoveries in modern chemistry. Vauquelin was not present the first time I was there; but two sisters of Fourcroy, who live at Vauquelin's were so good as to show me this beautiful laboratory. They seemed to be well informed of everything there, and told me, that they often assisted their brother and Vauquelin in chemical operations; but the old saying, that learned females are not always the handsomest and neatest, was verified in the persons of both those chemical ladies.

The cheerful, kind, and obliging Haüy resides at the Mineral School, as do also most of the inspectors and lecturers. On my first visit, I found him engaged in determining the specific gravity of a calcedon by means of Guyton's gravimeter. This instrument is an improved hydrometer. If my memory does not fail me, this instrument is described in Gren's *Physical Journal*, and is very much like the aerometer of Haüy and Nicholson. This instrument has several defects. In my opinion, the specific gravity of bodies can be determined with greater accuracy, by means of a good hydrostatic balance.

The figures of the crystallized bodies were investigated, and in some degree determined by Romé de l'Isle. Haüy has since thrown much light on the subject. He has contrived an instrument for measuring the angles of crystals, which consists of two small wires moveable about a fine pin. The one carries a finely graduated scale, or angle-measurer, and the other extends over the scale; so that when the angle of the crystal is included between those two wires, the scale measures a space equal to the vertical angle. There is also a mechanical contrivance, which fixes the limbs till the instrument be taken off the body, in order to observe the magnitude of the angle. Haüy has discovered a method of determining the forms and angles, not only of particular salts, but also of stones, earths, and metals, and he has exhibited beautiful specimens of this method, as practised on subjects in his own collection. He took a regular piece of quartz, and, from different examinations of the exact measurement of the space it occupied, immediately derived the crystalline structure of the body, which is two opposite pyramids, whose common base is a rectangle differing little from a square. A man resides in the school, whose employment is to cut in wood the different forms and figures of crystals under the direction of Haüy.

Thus has Haüy brought mineralogy, by the figures of its objects, and his own calculations under the dominion of geometry. He has published a work on[13] mineralogy . . . and Coquebert has edited his Instructions to the students at the School; but these works have not yet reached the hands of the booksellers. Haüy is at present engaged on an extensive and complete system of mineralogy, in which he gives the character of every mineral as depending: 1. On external appearances; 2. On the forms and figures of its crystals; and 3. On chemical analysis and synthesis.

Haüy had a pair of small instruments for ascertaining minute degrees of magnetism and electricity. On a small round stand is fixed a well turned steel pin, an inch in height; and on this point or pin, is placed a moveable needle, repeatedly magnetised, and about two inches in length. He took different iron ores, which he

[13]"Journal of"—J. J.

arranged in a straight and apparently exact line, coinciding with the required polar line, making thereby, in fact, a collection of several small and weak magnets. When, for instance, the crystallized iron ore from Norway is used, in order to ascertain the small degree of electricity, a brass needle is put instead of the magnetized one. Before the needle and in a line with it, must be laid a piece of rosin barely electrified, and over it a stick of sealing wax. This is a negative electric; and, since every body which comes into a negatively electric atmosphere, becomes positively electric, the needle will be positive. Haüy shewed me very clearly that different kinds of stones, by being heated, become electric, showing the positive electricity by repelling the needle, and the negative by attracting it.[14]

Before I close this account of the Mineralogical School, I must observe, that this was to be only a theoretical school; that a practical school was to be established in a mining district; and that Giromagny, in the department of the Upper Rhine, was the place fixed upon; but this school has not been yet organized. The Mineralogical School at Paris has, in the meantime, been regulated and modified in such a manner, that it unites the objects of a theoretical and a practical school. There is regularly[15] published, by the professors of this institution, a very important work, entitled *Journal des Mines.*

Ecole de géographes

The Geographical School, under the direction and management of Prony, is attended by twenty students, who are taken from among such as have completed their course in the Polytechnic. Here they are taught every branch which relates to the measurement of land, the drawing of maps, and such problems

[14]The phenomenon of pyroelectricity had been studied by Aepinus in 1756, when he showed that tourmaline when heated possessed the electrical property of attracting and repelling light bodies. Haüy in 1787 and 1791 published major studies on pyroelectricity in several other minerals. He discovered the relationship between pyroelectric state and the lack of symmetry in certain crystals.
[15]"annually"—J. J.

as occur in trigonometry, astronomy, and mathematical geography. The students then proceed to a finished method of planning, and to make astronomical observations, which they apply to the determination of latitudes, longitudes, and meridian lines; and to the construction of geographical maps. The students undergo an examination in all these subjects, before they quit the school.

Ecole des ingénieurs des vaisseaux

The School for Naval Architecture is in the Rue Dominique, No. 1016. This institution existed at Paris long before the Revolution, and the managers admitted whomsoever they pleased. But this custom has been altered, and no one can be now admitted, who has not first studied at the Polytechnic. The students have each 1,500 francs per annum, and are taught mechanics, hydrostatics and hydraulics, as far as they relate to naval affairs; to draw plans and sections of ships of war, to give an exact account of their expense, and even to superintend the building of them. This school is obliged to admit annually, from among the private merchants, five pupils who are also instructed in naval architecture.

The students had formerly their drawing lessons at the Louvre, where the National Institute is now held, and where is still to be seen a collection of naval models. But in lieu of this place, they have at present a General Marine Depository, in Rue Vendôme. This situation is a much finer one, and the present depository contains a more elegant and useful collection of naval models and drawings. Borda and Dudin are directors of this school, and Laplace is the examiner, Filz is professor of mathematics, and Pomet of architecture. Deparcieux lectures on physics, and Fourcroy on chemistry, and Daubenton teaches the students drawing. I must confess that there is too much of physics and chemistry delivered here, when it is considered that the students are all from twenty to twenty-four years of age, and have attended to both subjects, not only in the Central School, but also in the Polytechnic; so that they must have necessarily acquired sufficient knowledge of them, and their studies might now be more advantageously conducted, than in attending to these sciences a third time.

Ecoles d'artillerie

The great preparatory school, for students in artillery, is at Châlons sur Marne. The directors of the school are a *chef de brigade* and a *chef de bataillon*. There always reside in this school two captains of artillery, a lecturer on physics and chemistry, two on mathematics, two on fortification, and a drawing master. Laplace is the present examiner. Those students who wish to enter into the artillery service, are obliged to study for at least two years at the Polytechnic. They then enter, after a close examination, into the regiment of artillery; but must still prosecute a necessary and extensive study of drawings, fortifications, and the warlike preparations connected with the artillery service, in the School of Application belonging to their respective regiments.

The following is a list of the Schools of Application, with the places where they are established. The first school is at La Fère, the second at Besançon, the third at Grenoble, the fourth at Metz, the fifth at Strasbourg, the sixth at Douai, and the seventh at Auxonne. In each are a teacher of the mathematics, a private teacher and a drawing master; and every school is placed under the inspection of a general of the brigade of the artillery. There are to be two more artillery schools, one at Toulouse and the other at Rennes, but they are not yet organized. The artillery school at Châlons is to continue in its present state till peace be concluded, when it will undergo new regulation, and it is supposed that the preparatory school will be removed to Paris.

Ecole de fortification

The Fortification School, with which that of the Miners is united, is at Metz, and established in the *ci-devant* abbey of St Arnould. The number of pupils is not to exceed twenty: they must be all taken from the Polytechnic, and, when examined and admitted into this school, are immediately made second lieutenants, and receive the pay due to that rank. Here they are taught to apply their theoretical knowledge in founding, and actually building, works of defence and fortification, in mining and countermining, in defending and besieging places, in drawing military plans and maps, and in every art and science which belongs to the business of an engineer, both in fortified places and in the field. The School at Metz is under the inspection of a general and two *chefs de brigade*, who all three belong to the department of fortification.

Marine Schools

The Marine Schools are established at Brest, Toulon, and Rochefort. The students are admitted into them after previous examination in arithmetic, algebra, geometry, statics, and navigation. I have not been able to procure any certain account of the numbers of scholars, or of their plan of study. It was proposed and ordered, that a corvette should be annually equipped for different expeditions, with students on board, who should be instructed in rigging and unrigging a vessel, and taught by practical knowledge and experience, everything which belongs to the duty of a mariner. But the war has in some measure obstructed the execution of this commendable decree.

The Navigation Schools are intended for teaching mathematics and hydrography both to officers in the navy, and those of merchant ships. By a decree of 30 vendémiaire year IV "these schools were to continue in the state in which they had hitherto been," and the Marine Minister was directed to establish two other schools, one at Morlaix and the other at Arles.

Medical education

During the revolutionary period the universities came under attack as centers of privilege and decadence. They had functioned independently of the government, thus conflicting with the new view of education as the concern of the state. The Paris Faculty of Medicine was suppressed together with all faculties of theology, medicine, law, and arts throughout France by decree of the Convention on 15 September 1793. It did not take long for the government, however, to realize that, whatever else might be superfluous, France could hardly dispense with the teaching of medicine. It was all the more urgent because of the need for physicians and surgeons for the army. On 4 December 1794, the Convention issued a decree establishing three medical schools, one in Paris and the others at Strasbourg and Montpellier.

A new approach to healing was symbolized by the abolition of the terms *médecine* for the subject and *médecin* for the practitioner. The new schools were originally to be *écoles de santé* and a doctor was to be called *officier de santé*, thus rejecting the connotation of privilege of the medical profession before the Revolution. The term *santé* also connoted a concern for all the healing arts (medicine, surgery, and pharmacy) rather than medicine exclusively. Fourcroy, who had proposed the foundation of the *écoles de santé*, had some influence in establishing a *practical* curriculum: "Read little, see much and do much—this will be the basis of the new teaching." A new chapter in the history of medicine was opened with the predominantly clinical instruction of the Paris school. The use of the buildings of the former school of surgery no doubt helped to emphasize the practical approach to medicine. The study of medicine was now combined with that of surgery, with unusual stress on the latter. A critical comment of an English physician on French medical education may be quoted:

> The plan of giving lectures to the pupils free of expense is a bad one as the teachers may not always exert their talents sufficiently to explain the salutary

> art. The surgeons here operate very well but appear ignorant of the medical treatment of their patients and numbers die from this very cause. . . .[16]

To make room for the Ecole de santé in the rue des Cordeliers, a neighboring church was demolished and new buildings were constructed. An excellent library of 15,000 volumes had been established by 1798 and was used fully by the students. Fourcroy in his report to the Convention had stressed the importance of selecting the very best men to fill the chairs at the Ecole de santé. In this way France could demonstrate to the rest of the world the beneficial effects of her revolution. The men appointed in 1794 were certainly distinguished. Fourcroy himself was persuaded to accept the chair of medical and pharmaceutical chemistry. J. N. Hallé was appointed professor of hygiene. The treatment of internal clinical cases was in the hands of Corvisart, who became Napoleon's personal physician in 1805. Later appointments to the school included A. L. de Jussieu (natural history, appointed 5 April 1804), and Cabanis, who had taken an active part in the reform of medical education, also had an appointment (associate professor of legal medicine and the history of medicine, 28 May 1799–6 May 1808), although his bad health prevented him from carrying out his duties.

The intellectual advantages to be gained by the medical professors by the fact that each course of lectures only lasted six months did not escape the Austrian visitor, Joseph Frank, who studied the organization of the Ecole de Médecine when he was in Paris in 1803:

> If we add that the Professors are able to divide the work in each subject among themselves for half of the year, then it can be seen that the duties of a teaching

[16]E. J. Eyre, *Observations made at Paris during the Peace*, Bath, 1803, p. 279.

> chair need not prevent the professors of the *Ecole* from pursuing their private research, making themselves familiar with the literature of their own and foreign countries and thus continually increasing their own reputation by the publication of new works.[17]

The Collège de pharmacie had been set up under the *ancien régime* and it continued to function throughout most of the revolutionary turmoil. After the suppression of the Academies in 1793, the Paris pharmacists took the precaution of reconstituting themselves as a *société libre* in conformity with the law of 1 vendémiaire year IV (23 September 1795), which permitted such organizations. The Société took responsibility for the training of pharmacists and its Free School of Pharmacy was formally recognized on 22 May 1797 by the Directory.

From medicine and pharmacy, Bugge turns abruptly to technical drawing. The Ecole gratuite de dessin[18] had been founded in Paris in 1766 on the initiative of the painter Bachelier, who hoped that it would become a general technical school for industrial workers. By the ingenious grouping of the 1,500 boys into different sections and classes, they were all enabled to follow courses of four hours a week in the small space available. The course described here for 1798 differs little in principle from that established under the *ancien régime*.

The wording of the section on the Ecole de médecine differs slightly from the 1801 English edition, in particular the titles of the respective chairs of medicine.

Ecole de médecine

The Medical School is very beautifully situated in Rue des Cordeliers. It contains—a great number of excellent anatomical

[17] J. Frank, *Reise nach Paris, London, . . .*, 2nd ed., Vienna, 1816, 1,134.
[18] Bugge's juxtaposition of this technical school and the School of Medicine can be justified only on topographical grounds—they were both in the same street, the rue des Cordeliers, best known in the earlier history of the Revolution for its Club.

preparations, and imitations made of wax; a valuable collection of chirurgical instruments; some philosophical apparatus; a large library consisting of works on physiology, chemistry, anatomy, surgery, and medicine; a truly magnificent lecture room, or amphitheatre; and a beautiful chemical laboratory and reading room. The lecturers are two in number to each of the following divisions.

1.	Anatomy and physiology	[Chaussier and A. Dubois]
2.	Medical chemistry and pharmacy	[Fourcroy and Deyeux]
3.	Medical physics and hygiene	[J. N. Hallé and Pinel]
4.	External pathology	[Chopart and Percy]
5.	Internal pathology	[Doublet and Bourdier]
6.	Medical natural history	[Peyrilhe and Richard]
7.	Surgical operations	[Sabatier and Boyer]
8.	External clinic	[Desault and Boyer]
9.	Internal clinic	[Corvisart and Leclerc]
10.	Advanced clinic	[Pelletan and Lallemand]
11.	Obstetrics	[A. Le Roy and Baudelocque]
12.	Legal medicine and history of medicine	[Lassus and Mahon]

There are besides, a draughtsman and modeller in wax. A room is now being built for the library; in the place where the books now are, the anatomical preparations will be deposited, and more convenient apartments will be appointed for the chirurgical and philosophical instruments, and for the objects of the Materia Medica. This school is carried on with great industry, and the number of students amount to from one thousand to one thousand two hundred.

Ecole de pharmacie and the Military Hospital for Instruction

The body of apothecaries of Paris in the year 1777 were formed into a regular college. They have a laboratory and a botanic garden in the Rue de l'Arbalète, where lectures are publicly delivered on chemistry, pharmacy, botany, and natural history; and at the annual close of those lectures, prizes are bestowed on the most able and diligent students. In the fourth year of the Republic, the college formed itself into a free society, for the cultivation of the scientific pursuits connected with their profession, and admitted members from all the departments of France, and even from hostile countries. By a decree of the 3 prairial year V the Directory, in a message to the Minister of the Interior, approved and confirmed its system of public instruction, and gave it the name of the Free School of Pharmacy. In this school are two lecturers on pharmaceutical chemistry, together with an honorary professor and an assistant; two on pharmaceutical natural history, and the Materia Medica, with an assistant; and two on botany, with an assistant. The Free School of Pharmacy[19] consists at present of one hundred and twenty-three regular, and fifty-two corresponding members. A journal is published by this society, under the title of *Journal de Pharmacie*.

To the class of institutions for surgery and medicine, is to be annexed the Military Hospital for Instruction in Rue St. Jacques, not far from the National Observatory. It was formerly that well known and beautiful building Val de Grace.

According to the Programme, or account of the institutions, the following lectures were delivered there, in the year VII or from 22 September 1798 to 21 September 1799. The first, or winter course, consisted of: 1st, Anatomy, with physiological observations, by Huttier; 2dly, Internal pathology, by Chairon; 3dly, Practical medicine, and particularly clinical cases, by Gilbert; 4thly, Practical surgery, by Barbier; 5thly, Natural history, with reference to the Materia Medica and pharmacy, by Perinet. In the summer half year were explained, 1st, Pharmaceutical chemistry, by Brongniart; 2dly, Observations on gunshot wounds, by Dusouart; 3rdly, The diseases and setting of the bones, by Huttier; 4thly, Botany, by Barbier, who takes botanic excursions with the pupils.

[19]"Pharmaceutic School"—J. J.

Clinical lectures are read by all the six professors in medicine and surgery, with medical conferences and prescriptions in the morning; and, in the afternoon, consultations are held, on clinical diseases, in the amphitheatre. In the last decade of thermidor, or about the middle of August, a general examination of the students takes place, in order to confer prizes on such as have distinguished themselves by assiduity in the service of the hospital, or by attending the lectures and acquiring knowledge.

Ecole gratuite de dessin

The Free School for Drawing[20] is in Rue des Cordeliers. This patriotic institution was established thirty years ago, and was at first a private foundation for instructing in the principles of drawing, one thousand five hundred children, intended for artists or professors, but it is now rendered quite general. Every first, fourth and seventh day, of the decade, the students are taught arithmetic, practical geometry, statuary perspective and architecture; every second, fifth and eighth day, they paint men and animals; and every third, sixth and ninth day, they draw flowers and ornaments. Of this school there are two directors, who manage its concerns, a treasurer and five lecturers.

The National School for Architecture is at the Louvre, or National Palace for Arts and Sciences, and consists of a professor of geometry, who, at the same time, shows the application of that science to architecture, and of a professor, whose business it is to teach architecture in particular, with its subsidiary arts.

The Collège de France

The Collège de France (Collège royal) was founded by Francis I in 1530 as a center for "humanism" independent of the University of Paris. From about 1772 science had become increasingly prominent in the curriculum. The reform of this time in accordance with the ideas of the

[20]"Painting"—J. J.

Encyclopédie helps to explain why this institution underwent so little change after the Revolution. Perhaps the traditional democratic rights of the staff, who influenced both curriculum and appointments, was also a factor in allaying criticism. The Collège royal functioned until September 1793. On 21 November 1794, it was given a solemn reopening ceremony and on 13 July 1795, it was renamed the Collège de France. The staff fortunately resisted an attempt to deprive it of the chairs of medicine, natural history, chemistry, and anatomy, which seemed to some to belong more naturally to the Muséum d'histoire naturelle. The comprehensive curriculum of the Collège with work at a high level made it effectively the equivalent of a compact University of Paris, at least until the establishment of the Université de France by Napoleon in 1808, with the traditional faculties of medicine, law, theology, and arts (now called *lettres*), and the new faculty of science.

In 1805 there were no less than 1,500 students of all ages at the Collège de France, including many foreigners. The most popular courses then were those of Cuvier and Thenard on natural history and chemistry respectively; each course was attended by more than 200 students. Next in order of popularity were the courses in general physics, medicine, anatomy, Latin poetry, and French literature. Higher mathematics was taken by only about twenty students and the course on oriental languages was deserted.

The French College is situated in the Place de Cambray, close to Rue St. Jacques, and is a very ancient institution. Louis XII, and Louis XIII enlarged it, and either repaired or built new most of the present edifice. This college is the only institution of the kind which has not undergone some change during the Revolution. To give you an accurate idea of what is taught here, I shall subjoin a list of the lectures for this year, 1799, in the order communicated to me by Lalande.

Jerome de Lalande is inspector of the College and professor of astronomy. All the parts of astronomy, with their use in navigation, are explained by that distinguished master, or in his absence, by Michel Le François de Lalande.[21] Mauduit, professor of geometry, lectures on that science, and on trigonometry and algebra. Cousin, professor of theoretical[22] physics, lectures on the Analysis Infinitorum, with its applications, particularly to mechanics. Lefèvre-Gineau, professor of experimental philosophy, gives a complete course on that subject. Boerchart, professor of medicine, explains Stoll's aphorisms on fevers and feverish diseases. Portal, professor of anatomy, lectures on the causes and set of diseases. Darcet, professor of chemistry, explains the chemical analysis of different substances. Daubenton, professor of natural history, discourses on that science at the museum in the botanic garden.

Bouchard, professor of natural and popular laws, expounds political rights. Lévesque, professor of history and morality, delivers a course of lectures on the history of Greece, political, literary, and philosophical. Riviere, professor of the Hebrew and Syriac languages, expounds the text of the fourth and fifth books of Moses. Caussin, professor of the Arabic, teaches his pupils to read and write that language, and to translate Lockman's fables, and a part of Bidpai's moral and political work. Perille, professor of Turkish and Persian, or in his absence Silvestre de Sacy, explains the first principles of the Persian language, Hāfiz, and the Expedition of Diamis Beharistan. Bosquillon, professor of Greek, delivers philological commentaries on the *Prognosticon* of Hippocrates; and Gail, professor of the same language, expounds the *Corona* of Demosthenes. Dupuis, professor of Latin eloquence, and in his absence, Gueroult, interprets Tacitus' *De Oratoribus*. Delille, professor of poetry, or in his absence, Selis, explains the principles of poetry in general, and of epistles in particular; he expounds most of Horace's epistles, and compares that poet with Boileau, Rosseau, Voltaire and Pope.

[21] "Francis de Lalande"—J. J.
[22] "theoretic"—J. J.

Cournand, professor of French literature, treats of French literature, as compared with that of Greece and Rome, beginning with French fables. Each of the above named professors delivers four lectures in every decade.

It is evident, from the nature of this institution, and of the lectures delivered, that in France the utmost attention is still given to Grecian, Roman and Oriental literature. But as the youths are only initiated in the languages, at the Central Schools, the professors at the *Collège de France* are under the necessity of beginning with the first principles of language, and of course can make but a slow progress. Of these seventeen professors, six together with the inspector, live in the College. In the Central Schools also, several of the professors reside on the spot.

By an invitation of Lalande, I was present on 26 brumaire (16 November), at a public assembly of the *Collège de France*. A meeting of this kind is annually held at the commencement of the lectures. The auditorium of the College is a large and elegant hall, beautifully painted. The ceiling, in particular, is covered with this fine species of decoration. In the middle of the hall, is a very long table covered with cloth, about which are seats for the minister, professors, and visitors. On both sides of the table, there are forms for other hearers; but these are inconvenient, and rather too far distant, so that one cannot come sufficiently forwards. The hall was tolerably well illuminated; but such a strong and disagreeable draught passed through it, that the wax candles, both in the chandeliers and on the table, melted away before they were half burnt, and the meeting, near its close, was almost in the dark.

About seven o'clock came François de Neufchâteau, Minister for the Interior, attended by the professors of the college, and he took his seat at the head of the table, on the right of the inspector Lalande. The minister appeared in his public dress, and was accompanied by his adjutants, officers, and other attendants. Lalande opened the meeting with an oration, or rather a *praelectio*, in which he briefly enumerated the lectures on different branches of science, which were to be commenced. He then exhibited a few biographical sketches of some of the deceased lecturers of this college; and proceeded to give an account of his astronomical labours, and of the many thousands of telescopic stars which he had observed, in conjunction with his nephew, Michel Le François de Lalande. Much of his discourse was taken up with separate narrations, which unavoidably produced frequent and abrupt transitions. The history of astronomy cannot be unknown to those who read the *Connaissance des Temps*, much of the contents which had already appeared in Zach's *Geographical Ephemeris*: as an instance, we may mention the account that Dr. Burckhardt, who resided a year with Lalande as a student in astronomy, had calculated to a day, the orbit of a comet.

Dupuis, author of *Origine de Cultes*, recited some historical accounts of the Pelasgi, in which he introduced rather violent sallies against kings in general, and against all states which were not republics, interlarding, however, his philippic with some strong eulogia on Bonaparte, and several compliments to François de Neufchâteau. Most of the gentlemen who spoke offered to that minister some incense of commendation, which indeed the worthy man perfectly merited. But from what passed, one might see that the *captatio benevolentiæ* exists in republics, as well as in monarchical governments.

Bosquillon recited some aphorisms of Hippocrates; but his delivery was bad, and his voice so indistinct, that I could not comprehend the whole of his meaning. Delille recited part of a poem on Youth, a work on which he is at present engaged. The mathematician, Cousin, read a pleasing essay on Benevolence towards the Poor and Sick. He particularly wished that the Ladies would make themselves useful by visiting the sick, and even by attending the hospitals. Gail, professor of Greek Literature, recited a tract on the Spartan Republic, being part of a work which he intends to publish: he concluded with several translations from Anacreon. Gail's genius appeared to much better advantage, in historical and prose composition, than in poetry.

The whole hall was crowded with auditors, among whom was a party of ladies, who, by clapping of hands, assisted in applauding the speakers, and particularly Cousin, who introduced into his discourse some beautiful sketches, and admirable traits of the liberality and tenderness of the fair sex.

The French College has a collection of philosophical instruments, the greatest part which formerly belonged to the old institution. This college has also an observatory; but I must defer the description of it till another opportunity, when I shall give an account of the great national observatory, and others of less note, which I have visited at Paris.

The Muséum d'histoire naturelle

The Jardin du Roi was founded in 1635 by Louis XIII as a botanical garden to provide a supply of plants useful in medicine. In the course of the next 150 years the functions of the Jardin were gradually enlarged and it provided courses of instruction not only in botany but in chemistry and anatomy. The administration of the Jardin was the responsibility of a succession of *intendants*, of whom the most famous was Buffon. The Jardin is remarkable as the only scientific institution to continue to function throughout the Revolutionary period. Indeed in June 1793, just two months before the suppression of the Academies, the Jardin was expanded and given a new organization as the Muséum d'histoire naturelle. It was as if the disciples of Rousseau had shown favor toward natural history at the expense of the physical sciences. The cooperative attitude of the staff and their desire to reform their establishment from within did much to forestall outside criticism. Apart from the fact that the Jardin had friends in high places (for instance, Lakanal), it was open to all and its activities were associated in the popular mind with general utility. In the next few years the collections were considerably expanded by requisitioning and by numerous gifts. Although adjoining property was acquired by the Muséum, so many specimens of animals, vegetables, and minerals were accumulated that, as Bugge says, "There is no room for more objects without additional buildings." Further buildings were added under the Consulate and Empire, so that the Muséum became one of the sights of Paris and was inspected by many visitors without the least pretension to any knowledge of the sciences represented.

The constitution of 10 June 1793 provided a total of twelve professorships. Particularly significant for the history of science was the establishment of the study of animal as well as human anatomy. This institutional structure led Cuvier to make a significant contribution to comparative anatomy. A new career began for Lamarck, when, at the age of nearly fifty, he was asked to turn from the study of botany to that of lower animals.

Although the botanical garden was at some distance from the center of Paris, many visitors looked on it rather as a beautiful park with the scientific collections of the Muséum as an adjunct. At least one comprehensive work was published specifically with the object of explaining the contents of the institution to the layman:

> J. B. Pujoulx, *Promenades au Jardin des Plantes, à la Ménagérie et dans les Galeries du Muséum d'Histoire Naturelle... Contenant des Notions claires et à la portée des Gens du Monde...* (2vols., 12mo., 1803–1804).

Another comprehensive work, stressing the constitution of the Muséum, was published by a German visitor in 1802.[23]

Pujoulx in the introduction to his book gives an idea of the size of the collections in 1803:

> The *Jardin des Plantes*, the orangeries and the glass-houses now contain more than 6,000 species of plants,... the gallery devoted to stuffed quadrupeds offers a collection of more than 400 specimens,... the collection of birds which is the most brilliant in Europe contains more than 2,000 of them; ... reptiles and fish fill more than 600 jars, ... shell fish, insects, worms, sea-urchins and madrepores form a collection

[23]Gotthelf Fischer, *Das Nationalmuseum der Naturgeschichte zu Paris*, Frankfurt am Main, 1802.

> of more than 9,000 species, . . . there are more than 20,000 specimens of minerals, fossils, wood and fruits.[24]

The book then proposes and describes four "promenades" through the gardens.

Yorke was one of those who found that the scientific objectives of the Jardin des plantes overruled aesthetic considerations:

> The botanical garden is . . . in a better state than when I last saw it in 1792 but it is neither well laid out nor is the aspect pleasing to the eye. I know nothing to which it can better be compared than a Dutch kitchen garden. Its chief merit (and that is considerable) consists in the correct classification of the different plants and trees; and in its being the nursery whence all the central schools in the departments are to be furnished with seeds and plants . . . . The garden is about 2,000 feet long and 700 wide, divided into three alleys.[25]

Although lectures had been given there before the Revolution, a new importance was attached to teaching when it became the Muséum, so that not only could its collections be admired but it could be described as "one of the first establishments of instruction existing in Europe."[26]

Benzenberg visited Paris in 1804 and took full advantage of the lectures offered at the Muséum:

> The lectures are all public. The professors are paid by the government. . . . The majority live in the *Jardin des plantes* near the large galleries. I attend the lectures of Geoffroy [Saint-Hilaire] on mammals; he lectures from 9 to 10 in the great gallery. From 10 to 11 I listen to Haüy on mineralogy . . . . In the afternoon from 3 to about 5 Fourcroy lectures on chemistry in the amphitheater on Wednesdays and Saturdays. Vauquelin lectures on

[24] Pujoulx, *Promenades au Jardin des Plantes*, I, 10.
[25] Yorke, *Letters from France*, I, 236.
[26] [Blagdon], *Paris as it was*, II, 449.

> technical chemistry at the same time on Fridays. I attend the lectures of both. . . . One spends the hours between lectures either in the library or in the galleries with the mineral collection or the birds or the quadrupeds. . . . The botanical garden is so large and rich in so many things that one always finds a pleasant occupation even if one is there for the whole day.[27]

On days when there were lectures Benzenberg went to the Muséum at 8 in the morning and would only return to his lodgings in the evening. Having formally enrolled in Haüy's course, the German student had a student card which enabled him to gain entry to the Muséum on those days when it was closed to the public. Haüy's lectures, although described as "mineralogy," also included discussions of related sciences. Thus when the topic was lead, the use of lead oxide in making flint glass was described, and the introduction of zinc led to a discussion of the voltaic pile.

The National Museum of Natural History was formerly called *Jardin du roi*; but received its present name by a decree of the National Convention of 10 June 1793. One end of it extends to the Seine: it consists of a botanic garden, library for natural history, a menagerie, or collection of foreign animals, and an amphitheatre, or lecture room.

The botanic garden which belongs to it is three hundred and twenty *toises*, or fathoms, long, and one hundred and ten in breadth. It is partitioned lengthways, that is, from its entrance down towards the Seine, by three very fine alleys; and intersected across by various others, which terminate in the public promenades, or walks. The different square divisions thus formed, are used for plantations, and are at present all enclosed with rail-work. The green-house and orangerie were formerly in pretty good order, and separated into rooms and spaces. But

[27] J. F. Benzenberg, *Briefe, geschrieben auf einer Reise nach Paris*, Dortmund, 1805, I, 123–124.

a new green-house and orangerie are now additionally erected, and they are very conveniently disposed. Here is a great abundance of foreign plants and trees, and from hence all the botanic gardens of the Central Schools are supplied with seeds, and with trees, as soon as they can be transplanted. From the same highly cultivated spot, the cultivators of land can procure economic and nursery trees, and even the indigent poor can obtain plants, when they can be spared.

Captain Baudouin, in his travels into different parts of the world, had collected a great variety of natural curiosities, and presented the whole to the nation, on condition that he should be furnished with a ship to convey them to France. The English Government consented that this ship should perform her voyage without molestation. Meanwhile the English had taken possession of the island of Trinidad, where this extensive and famous collection had been left. When Captain Baudouin arrived at Trinidad, in order to bring away his collection, the English would not give it up, on pretence that their Government had consented to the safety of the expedition by sea, and not by land. However, this and the former expeditions were not altogether fruitless; for Baudouin has brought into the botanic garden about one thousand different kinds of live plants, besides assortments of seeds, and a considerable herbarium.

The gallery for natural history is a building situated on the right hand, as you enter the botanic garden from the street. On the second floor of this building are four large apartments, where fishes, birds, shells, insects, minerals, earths and stones are deposited on shelves, furnished with glass fronts. The inner apartment is allotted to vegetables, and contains specimens of trees, together with the herbarium of Tournefort.

Vaillant presented to the Museum a part of his birds. But several persons, who had certain knowledge of the fact, assured me, that Vaillant reserved for himself the most singular and curious.

The gallery is open to the public the first, fourth, and seventh days of every decade, when it is crowded by all sorts of people, who come there not for instruction, but merely to view the place, by way of amusement. A certain number of veterans and invalids are then stationed in different places about the rooms, in order to see that the drawers are not broken open, or the curiosities in any manner injured or destroyed. Before this regulation took place, a diamond was stolen from thence, in the time of the Revolution. Every second, third, fifth, sixth, eight and ninth days of the decade, this gallery is open for such only as are desirous of studying natural history.

The excellent Lacépède, who is not less kind and obliging than eminent for erudition, gave me a letter to Lucas, keeper of the gallery, who, with great civility, shewed me everything that was curious and remarkable in this museum, and particularly the collection of quadrupeds, which is never exhibited to the public. Here I had a second view of some singular objects, which I had seen at the Hague one and twenty years before, in the Stadtholder's collection, such as the sea-horse, zebra, elephant, orang-outang, and a variety of monkeys. There are likewise to be seen in this museum, a lion, a tiger, a leopard, an uncommonly large dog from the Pyrenees, and a fine skeleton of a camelopard, whose height from his forefoot to the top of his crown is sixteen feet.

All these and many other quadrupeds, and some large birds, are exhibited to view in an apartment on the third floor, or rather on a part of the garret formed into an apartment. The remaining part of the floor has the appearance of a large hall; above are sky-lights, and on each side are dens for wild beasts.

The Directorial Palace, formerly *Palais de Luxembourg*, is arranged with the utmost conveniency and grandeur, and is now the residence of the Five Directors. All the window glass which the great factory at Paris had made for some time, was ordered for that building. But the Directory had so much respect for science, as to part with whatever glass was wanted for the cases of the gallery for natural history; so that this large hall was soon fitted up with all the state and magnificence in which it now appears. In the building where Cuvier now resides, is a choice collection of skeletons of men, quadrupeds and birds.

It must be observed, that as natural history and botany are

almost unlimited, the descriptions which I give of the natural treasures of the garden, green-house, and gallery, cannot but be superficial. . . .

The library, which is on the second floor, by the side of the gallery, contains from nine to ten thousand volumes, relating to botany and several other branches of natural history. On the walls are hung several very masterly paintings of plants and animals, executed by the students at the museum. All remarkable plants and animals are drawn on vellum paper, and laid up in common bindings: the number of parcels of this kind is very considerable. This library was formed in the time of Louis XV, and has been continually increasing. It is open every day, excepting decade days, from eleven till two o'clock.

The menagerie seems to be separated into two parts, the one for mild animals, and the other for the wild and ferocious. Between two of the long alleys, on the right hand, as one enters from the city, and about the middle of the garden, are enclosures of very fine railing, within which are the mild animals, such as camels, dromedaries, African oxen, East Indian deer, several kinds of sheep, Angora goats. &c., &c. some of which have even propagated their species in this garden. The other part of the menagerie is for ferocious animals, which are kept on the left hand side, in a low building with different apartments. Here are a lion, four lionesses, a white bear, and several Alpine bears (which formerly had free possession of the state den of Berne, but now inhabit that of Paris), a wolf, an African porcupine, sea bears, &c. It is remarkable that there is a dog here which continually lives in company with a young lioness.

In a large cage are contained different birds of prey: such as eagles, griffins, hawks, storks, &c. On the left hand, on entering the garden, and behind the green house, are several buildings. In one of them are two grey elephants from Holland; and in another, two ostriches, a cassowary, and some antelopes. In other parts of the garden, are enclosures for land and sea fowls, and three ponds of spring water for fishes.

The amphitheatre is a remarkable building, which faces the garden on the left side. This lecture room indeed is an amphitheatre in its true acceptance, that is, the forms are all constructed in semicircles, and rise regularly one above another. At the centre below stands the lecturer. I attended a lecture on chemistry, delivered in this amphitheatre, by Brongniart. It was difficult to hear and understand him; but I cannot say with certainty, whether the cause was to be looked for in the voice of the speaker, or in the construction of the building. I was, however, rather inclined to ascribe it to the latter, as the voice must necessarily be confused by reverberation. In the same building there is a chemical laboratory.

The superintendents of this museum are [A. L. de] Jussieu, who is principal director and professor of rural botany, and the following managers, viz. Daubenton, professor of mineralogy; Fourcroy, of general chemistry; Brongniart, of technical chemistry; Desfontaines, of botany; Geoffroy [Saint-Hilaire], of the zoology of quadrupeds and birds; Lacépède, of the zoology of reptiles and fishes; Lamarck, of the zoology of insects, worms, and testaceous animals; Portal, of human anatomy; Mertrud, of the anatomy of animals; Thouin, of gardening; Faujas, of geology; Vanspaendonck, of ichnography, and who also teaches the students to take sketches of animal and vegetable objects, &c. Cuvier is adjunct in the anatomy of animals. All these professors deliver their public lectures in rotation, and in the summer months only. The other officers consist of a principal and sub-librarians, two keepers of the gallery for natural history, a gardener and a secretary. The assistant professors are, one in mineralogy, two in zoology, and one in botany. Two Captains constantly keep guard with their veterans at the museum. Most of the professors and officers have a free residence in buildings belonging to this museum.

There formerly stood in the library of the museum a statue of the celebrated Buffon, of Parian marble, and as large as life. During the Jacobin government, it was taken down, but preserved from damage. It is said that it will be restored to the former honourable situation, deservedly due to the inanimate representative of Buffon, whom the French have generally named the second Pliny.

Just below the entrance from the city into the botanical garden, and on the left hand, there is to be seen a plantation of trees and shrubs, which rise up to a considerable height, and have a beautiful appearance. In this fine grove formerly stood, under a noble cedar of Lebanon, a marble bust of Linnaeus, the Swedish nauturalist, and the inventor and founder of the modern system of natural history. This bust was destroyed, at the time when the *peuple souverain* amused themselves with spreading ruin and devastation. The cedar of Lebanon, either by a cannon ball or some other violence, then lost its majestic top. Those vandals destroyed every memorial and monument, without any discrimination whatever. They even demolished the tombs, and dug up the bodies, of the most meritorious of their countrymen; not exempting that of the great Turenne himself, who had been, more than once, the deliverer of France. His sacred remains, in which was still visible the wound of the cannon ball by which he fell, in the service of his country, were treated by those barbarians in the most inhuman and contemptible manner. The most mortal part of that great General lay in the museum, shamefully exposed among the skeletons of quadrupeds and birds; till it was removed by the orders of François de Neufchâteau, and then placed in an apartment of the amphitheatre, where it is set upright in a glass case.

Before I take my leave of the Museum for Natural History, I must observe, that it contains a great number of chests still unpacked, which are full of curious objects brought hither from conquered countries. I have been told by men, who had every opportunity of being well informed, that those chests inclose a collection as interesting and extensive as that already deposited in the museum, in which there is no room for more objects without additional buildings.

Chapter 2

THE NATIONAL INSTITUTE

The Académie royale des sciences had been suppressed on 8 August 1793 together with the other academies on the principle that there were to be no privileged corporations in the new Republic. The following year, however, the desirability was felt of bringing together the most eminent men in all branches of knowledge. In accordance with the principles of the great *Encyclopédie*, knowledge was not only the key to salvation but it had an essential unity and this was reflected in the foundation of a single institute, divided into three sections or "classes." As in the old academies, the membership of the Institute was strictly limited, thus providing a contrast with the situation in England. Whereas the members of the Royal Society were very numerous and among them competent men of science were then in a minority, the membership of the First Class of the Institute was limited to six of the leading French representatives of each separate science.

The very restricted membership of the First Class made the honor of belonging all the greater. Only in exceptional cases were young men able to become members. Membership was more often a reward after many years of research. Although members received a salary of 1,500 francs, this was equivalent to no more than what a graduate student might receive today. The comment of the British man of science, Thomas Young, was that such a salary, although "sufficient to induce men of small fortune and moderate wishes to devote their attention to science [is] by no means calculated to call the most brilliant powers into the strongest action; and a society so constituted is more likely to do a great deal tolerably, than a little admirably."[1]

It is clear that the salary of a member of the Institute was hardly substantial and one might ask why so much importance was attached to membership. The answer is that it was a great honor. But it was also the stepping stone to other positions, which often carried substantial salaries. Yorke expresses this in the following comment:

> These members of the Institute burrow everywhere; it is a matter of no small importance to obtain a seat in their hall, for it is the ante-chamber to wealth, fame and power.[2]

In particular, members had the first choice of appointments as consultants, as administrators where technical knowledge was required, and in specialized teaching posts.

Although those belonging to the Institute received salaries from the state, it would be misleading to think of them simply as civil servants. Over-all responsibility for the Institute was in the hands of the Minister of the Interior, yet even he was unable to direct their activities. At one of the meetings attended by Bugge (p.83), the Minister had tried to direct particular members of the First Class to examine a certain project. The choice of members of committees was a jealously guarded privilege and a polite rebuff was sent to the Minister.

It would be too idealistic to suppose that election to the Institute was based simply on merit. Yet, despite the intrigues which Pinkerton hints at (examples of which have now been documented),[3] there was no doubt of the great distinction of the majority of the members of the First

[1] *Quarterly Review*, May 1810.
[2] Yorke, *Letters from France*, II, 15.
[3] M. P. Crosland, *The Society of Arcueil*, London, 1967, pp. 161–168.

Class. Pinkerton's remarks in the following passage reflect both the realities of academic politics and the area of the Institute which was most distinguished:

> In all these Classes, as happens in such institutions, men of superior talents are mingled with one half or one third part of mere quacks, who have usurped a ridiculous reputation by low intrigues, and by taking advantages of particular times and circumstances. Still the mass of science and the freedom of enquiry are so predominant that the Institute may be regarded as a grand focus of illumination, particularly in natural philosophy and chemistry.[4]

Among German visitors, Reichardt provides an interesting picture of the procedure at a meeting of the First Class, thus presenting an aspect lacking from the formal minutes:

> Lalande had the kindness to introduce me to a meeting of the *Institut National* and to introduce me in the most flattering terms to the president, the Minister, Chaptal. . . . He [Chaptal] intervened several times in discussions to rectify with finesse and precision opinions expressed by speakers who became too animated. For the meeting did not have that calm that distinguishes our academic meetings on the other side of the Rhine during which one member generally reads a dissertation while the rest of his colleagues leaf through books, read papers or gossip among themselves. At the beginning of the meeting, the secretary, Lacroix, very active though lame, had read the minutes of the previous meeting. Then he presented the manuscript memoirs sent by foreign scientists to the Institute, at the same time making known the wishes and explanations contained in the accompanying letters and other members did likewise. The replies to be addressed to correspondents

[4] J. Pinkerton, *Recollections of Paris in the Years 1802–3-4-5*, London, 1806, I, 212–13.

were discussed and in this connection quite a few witty, lively and even biting comments were made.[5]

Benzenberg, in contrast, considered the reading of memoirs at the meetings of the First Class rather boring; yet he made a point of attending them as regularly as possible. He explains that he wished to avail himself of the opportunity of seeing so many famous men of science.[6]

Louis XIV and his minister Colbert, were both favourable to the sciences. Seeing their happy influence on navigation, arts, manufactures, and trade, they encouraged and patronized the cultivators of science and useful arts. In order to promote agriculture, and extend scientific inquiries, Louis XIV founded the Academy of Sciences [1666], which comprehended mathematics in all their branches, physics, natural history, chemistry, and medicine: he also established the Academy of Inscriptions and *Belles Lettres* [1663], the Academy of Inscriptions and Medals [1701], the Academy of Surgery [1731],[7] and the Academy of Architecture [1671]. These academies, as appears from their memoirs, have always consisted of able and skilful men, who have thrown new light on the arts and sciences, in their writings, and have enriched them by numerous and important discoveries. At the commencement of the Revolution, the Academy of Sciences in particular, included some of the greatest men in Europe, in their respective departments. To be convinced of this, we need only name the mathematicians Lagrange and Laplace; the chemists Lavoisier and Fourcroy; the natural historians and mineralogists Daubenton, Lacépède, and Haüy; the astronomers Lalande, Messier, and Delambre; not to mention many others who have contributed more or less to the extension of scientific inquiries.

During the Revolution, all preceding monarchical institutions

[5] Reichardt, *Un hiver à Paris*, pp. 77–78.
[6] Benzenberg, *Briefe*, I, 222.
[7] Bugge's history is at fault when he attributes the founding of the Academy of Surgery to Louis XIV, who died in 1715. It was established under the patronage of Louis XV.

underwent a change, and even the free temples of the sciences were subverted. Upon their ruins, was founded the National Institute, which not only comprehends all the branches into which the academies of sciences, and of the *belles lettres*, were formerly subdivided, but also includes logic, morals, and politics.

The decree, which established the National Institute, was passed on 3 brumaire year IV or 24 October 1795. According to this decree, the Institute belongs to the whole Republic; but is to be situated in Paris. Its object is, to extend the limits of the arts and sciences, by discoveries and inquiries, and by corresponding with learned societies in foreign countries. By the resolution of the Directory, the Institute is to undertake and promote such scientific labours as conduce to the general utility and honour of the Republic. It consists of 144 members, residing in Paris, and of an equal number in other parts of the Republic, and it may additionally admit eight foreign associates; but they have not yet been chosen. The National Institute consists of three Classes: the First, or mathematical and physical Class, is divided into ten sections, each of which has six members.[8]

1st section. *Mathematics*; Lagrange, Laplace, Borda, Bossut, Legendre, and Delambre.

2nd section. *Mechanics*; Monge, Prony, Leroy, Perier, [Bonaparte,] [Berthoud].

3rd section. *Astronomy*; Lalande, Méchain, [Bory,] [Le Monnier,] Messier, Jeaurat.

4th section. *Experimental physics*; Charles, [Cousin], Brisson, Coulomb, [Rochon], Lefèvre [-Gineau].

5th section. *Chemistry*; Guyton Morveau, Berthollet, Fourcroy, Vauquelin, [Deyeux,] [Chaptal].

6th section. *Natural history and mineralogy*; Darcet, Haüy, Dolomieu, [Desmarest,] [Duhamel,] [Lelièvre].

7th section. *Botany and plant physiology*; Lamarck, [Desfontaines,] Adanson, Jussieu, L'Héritier, [Ventenat].

8th section. *Anatomy and zoology*; Daubenton, Lacépède, [Tenon,] Cuvier, [Broussonet,] [Richard].

[8]As Bugge's list is incomplete, several names have been added.

9th section. *Medicine and surgery*; Dess[ess]arts, Sabatier, Portal, [Hallé], [Pelletan], Lassus.

10th section. *Rural economy and veterinary art*; Thouin, [Gilbert,] [Tessier], Cels, Parmentier, [Huzard].

There are, in all, in this Class, sixty members at Paris, and an equal number in the departments, where they are also divided into ten sections, each consisting of six members.

The Second Class comprehends moral and political science, and is divided into six sections, each consisting of six members; in all thirty-six members, and as many in the departments.

1st section. The analysis of sensations and ideas.—2nd section. Morals.—3rd section. Civil society and laws.—4th section. Government.—5th section. History.—6th section. Geography.

The Third Class is occupied with literature and the fine arts, and is divided into eight sections, each six members; in all forty-eight members in Paris, and as many in the departments.

1st section. Language or grammar.—2nd section. Ancient languages.—3rd section. Poetry.—4th section. Antiquities and monuments.—5th section. Painting.—6th section. Sculpture.—7th section. Architecture.—8th section. Music and Declamation.

Every Class meets twice in every decade; the First Class on the first and sixth days; the Second on the second and seventh days; and the Third on the third and eighth days. Each Class has its president and two secretaries, who are elected by the Class they belong to, and continue in office for six months and twelve months respectively.

On the 5th day of the first decade, in every month, the three Classes unite, and hold a general meeting, to deliberate on such affairs as relate to the general interests of the Institute. The oldest of the three presidents of the Classes then takes the chair, and acts as president of the whole Institute.

The National Institute has four public quarterly meetings; namely, on the 15th of the months of vendémiaire, nivôse, germinal, and messidor. Each Class annually proposes two prize questions; and in these general meetings, the answers are made public, and the prizes distributed. The united sections of painting, sculpture, and architecture elect the pupils who, at the expense of the Republic, are to travel to Rome, and to reside at the national palace, in order to study the fine arts. By virtue of a decree of 3 brumaire year IV, the Institute should likewise elect twenty young men, to travel in France and foreign countries, for the purpose of studying rural economy. Six members of the Institute itself, are also to travel at the public expense, in order to collect information, and to acquire experience in the different sciences. But I do not apprehend that any of these scientific expeditions have been performed; war and the want of money having probably obstructed these very useful undertakings.

It may be presumed, that the members I have named, in the several sections of the mathematical and physical Classes, are the most celebrated and eminent men in these scientific departments. The two other Classes are also composed of members equally respectable; and, upon the whole, it cannot be denied, that the National Institute of France is the first learned body in Europe.

Among the many pleasures I have derived from my travels, I account it the greatest that I have become personally acquainted, and frequently conversed, with so many excellent men, all eminent in their respective pursuits. I may particularly mention Lagrange, Laplace, Borda, Bossut, Legendre, Delambre, Prony, Perier, Lalande, Méchain, Messier, Jeaurat, Charles, Brisson, Coulomb, Lefèvre-Gineau, Fourcroy, Vauquelin, Darcet, Haüy, Lacépède, Cuvier, L'Héritier, and Grégoire. The remembrance of these excellent men will always be dear to me, and I shall ever thankfully acknowledge their friendship and civility. As soon as I arrived at Paris, I was presented by Mr. Secretary Dreyer, to Talleyrand de Perigord, the minister for foreign affairs, by whom I was introduced to François Neufchâteau, the minister for the interior, and by him again to the mathematical and physical Class, whose president at that time was Bossut, well known for his many excellent mathematical writings.

This cheerful, good man received me with much civility and friendship; he informed me who were the French commissioners for weights and measures, and that the Institute had resolved that the foreign commissioners, during their stay in Paris, should be considered as members of the Institute, and have free admission to every particular Class, and to their general and solemn meetings. He then delivered to me an oval printed card, inscribed round one side *République Française*: in the middle *Citoyen Bugge, Membre et Commissaire de l'Institut National des Sciences et des Arts*: and it was signed[9] *Cels, President de la Commission des Fonds, et de la Bibliothèque*. On the other side were printed the words, *Le Citoyen Bugge, Commissaire des poids et mesures, envoyé de Dannemark.*

The apartments of the Institute are on the first floor of the *ci-devant Louvre*, now called the *Palais National des Sciences et des Arts*. At the entrance is an elegant anti-chamber, through which one enters the hall of the Institute, which is oblong, lighted by windows in each end, and hung with tapestry. Small tables covered with green cloth are placed parallel with the walls and windows. In the middle of one of the longest sides, is the chair of the president, and his two secretaries are seated one on each hand of him. Straight before the president, in a rectangular space, is a table where those who have anything to read usually stand, particularly if they be not members of the Institute. Within this space, a table was placed for the foreign commissioners for weights and measures. The length of the hall is sufficient to admit twenty-six persons to sit at each of the longest sides of the tables, and about ten may be seated at each end, besides benches for strangers adjoining the wall and windows. On one side of this great hall, is a smaller apartment for the reception of the communications of correspondents. The library, in three large apartments, contains about 16,000 volumes, including the transactions and memoirs of the former French academies, and of foreign scientific establishments and literary societies. The Institute has also an apartment for the secretary and his assistant; and a large room for a collection of machines and models, wherein are many pieces of mechanism which belonged to the

[9] "subscribed"—J. J.

old Academy of Sciences; and a great number of models of all kinds of ships; for this room was once used as a model-room for the students of naval architecture.

Since the establishment of the Institute, there have been deposited here more than twenty models of machines, intended to enable people to escape from the upper stories of buildings on fire. Of these models the descriptions and drawings of three, which were looked upon as the best, have been published under the title of *Rapport sur les moyens de sauver les personnes, renfermées dans les maisons incendiés*, by Prony, Coulomb, Peyre, Bougainville, Perier and Boullée.

In the garden of the *Palais Royal*, now the *Palais d'Egalité*, a building had been erected, one hundred fathoms in length, and chiefly of wood. In 1798 this building took fire, and during this violent and dangerous conflagration, some of those machines were produced and tried; but were consumed along with the building.

I have been several times at the meetings of the Second and Third Classes and those of the First, or mathematical and physical Class, I almost always attended. I shall give some cursory accounts of what passed in those meetings. The president, Bossut, conducted me to the National Institute on the 11 fructidor, or 28 August. The meeting began, as it generally does, at six o'clock, and continued till eight. It was opened, as usual, by reading an abstract of the proceedings in the meeting, which was this evening followed by a chemical dissertation on the analysis of saliva, with physical conclusions thence deduced. Portal, the physician, did not seem satisfied with it, but objected that it was impossible to reason on effects in the human body, from those which casually took place in glasses and retorts.[10] Chaptal read

[10]Although Lavoisier in 1791 had successfully explained respiration as a chemical process, the validity of such reductionism was often challenged. The note by Bugge of this intervention by Portal is an example of the additional information that may be obtained from an informed spectator, for such comments are rarely to be found in the official minutes of the Institute.

a method of producing from vegetables, a material which communicates to linen and woollen manufactures, a much more beautiful and durable yellow than the common one.

Dizé, a pupil of Darcet, read a treatise on light and caloric, wherein he attempted to prove that these principles are always united, and are only one and the same element in nature. He had mixed alkalies and acids, which by combination produced heat, and he had very often seen, in the dark, sparks emitted by the mixture. Laplace, who has extended the limits of several branches of science, and who often speaks before the Institute, with that order and clearness of thought which might be expected from so eminent a man, raised a doubt, whether this light might not be electric. He said he remembered to have made a similar experiment along with Lavoisier; and had advised the author to repeat it, and for the greater certainty, to insulate the vessels.

Maingon, a lieutenant in the navy, produced a new graphical method of ascertaining the observed distance between the sun and moon, or a fixed star and the moon, in order to find the refraction and parallax, in observations for finding the longitude at sea. This treatise was delivered to Borda and Lévèque of Nantes (known for his excellent book on navigation) for them to report their opinion of it to their Institute.

The Institute in its manner of debating, resembles the English Societies. Any individual who is inclined to speak, asks leave of the president, to whom he addresses his discourse, and every individual speaks in the order which his inclination suggests. Hence those debates are carried on with regularity, decorum, and mutual respect.

At the meeting of the 16 fructidor, or 2 September, a programme was read by François de Neufchâteau concerning the festival of the 18 fructidor. Chaptal read a memoir, in which he and other chemists were disposed to prove, that there was an essential difference between acetic and acetous acids: his experiments and proofs appeared to me to be very convincing. Baumé, who is still an advocate for the phlogistic system, raised several objections. Fourcroy spoke with his usual elegance and solidity, and supported Chaptal's propositions. Two petitions from Baumé and Sage were produced, praying for an augmentation of their scanty allowance.[11] In consequence of the opinion of the Institute, those petitions were sent to the Minister of the Interior, and both were recommended to his attention. A similar petition was read from one who mentioned his having travelled with the Abbé Haute Roche, and assisted him by his astronomical observations: but as none of the astronomers knew this man, or had heard anything of his abilities and labours, his petition was not recommended.

In a meeting of the National Institute of the 6 vendémiaire year VII, or 27 September 1798, the famous botanist [Antoine Laurent de] Jussieu, elected president in place of Bossut, took the chair. Jussieu having a very good voice, and a regular and distinct delivery, made an excellent president. The meetings, in the winter months, begin at half past five and close at half past seven. A letter was first produced from the Minister for the Interior, enclosing a plan for altering and improving the water-works at Marly, proposed by Bralle, the engineer. Prony and Coulomb having already offered plans for a similar improvement, the minister proposed, that they should be authorized to examine this project of Bralle. Laplace, Borda, and many more objected to this proposal, whilst others supported it. It was finally decreed, by a majority of voices, to write to the minister, that, if he requested the opinion of the mathematical and physical Class, he must permit them to elect their own committee; but, if he only wished to have the opinions of Prony and Coulomb concerning this plan, it should be sent to them accordingly.

Guyton Morveau read an extract of a work sent to the Institute, concerning an analysis of Spanish minerals. Chaptal read a very favourable account and report of Dizé's memoir on

[11] This provides evidence to confirm that the salary received by members of the Institute (1,500 francs per annum on average) was not sufficient for their full support. Most members had other sources of income, for instance, from teaching. Baumé and Sage, as older chemists who had opposed the new chemistry, may have been discriminated against for appointment to other remunerative posts. The new chemists, Guyton, Berthollet, and Fourcroy, were all in powerful positions on government committees.

the identity of light and caloric. Laplace read an account of the disruptions of the dykes near Doel in Flanders, which happened under circumstances causing the highest tides. It was new moon, near the equinox, and the moon was, at the same time, in perigee, or nearest the earth, and consequently acting with her greatest possible force.

At this meeting was exhibited a model of a new telegraph, calculated to make signals by numbers. There were three perpendicular poles.... The first pole may be allotted to units, the second to tens, and the third to hundreds; and to express the numbers of each, long pieces of board can be suspended on every pole. [This arrangement may be combined with a system of colored flags to indicate various powers of ten.] On this principle, a signal system was proposed to be calculated, and a protocol of signals to be formed, wherein certain numbers were to denote certain syllables, words, and meanings, according to which the signals of the telegraph were to be given and read.

This method is undoubtedly well contrived; but it appears to me, that the telegraphs now used at Paris, with two moveable arms, which stand as different angles, in order to signify different syllables, are more simple in structure, and expeditious in practice.

Fourcroy, in the last place, read an excellent account of a chemical analysis of *calculi* formed in the human bladder, undertaken by himself and Vauquelin. He had examined more than three hundred of those concretions, and found that they all consisted of the same component parts; but that they ought to be reduced into different classes. He mentioned one monstrous stone of the size of a melon. It is luxury to hear this enlightened master treat of his science, with so much zeal and precision.

At the meeting of 11 vendémiaire, or 2 October, several members read various extracts of memoirs communicated to the Institute; but none of any particular importance. Lévèque read, in his own and Borda's name, a report concerning Lieutenant Maingon's graphical method of ascertaining the distance between the sun and moon, in order to find the parallax. This report began with an historical relation of the first attempts made for determining the longitude. The whole was written with great ingenuity; but a little national partiality still prevailed, and the very great service which Dr. Maskelyne has rendered in this business were not thought worthy of notice. That able astronomer, by his *Mariner's Guide*, first contributed to promote and introduce the methods of distances among the English navigators; and first proposed the publication of the *Nautical Almanack*, and the requisite tables to be used with it, which have afforded infinite assistance in the calculation of the Longitude.

Lévèque took some cursory notice of the several methods for calculating the Longitude suggested by Lyons, Dunthorne, Maskelyne, and Borda. There appeared some degree of meanness in so often naming and commending Borda, who assisted in drawing up this report; yet I am not on that account disposed to depreciate, in any respect, the merit of Borda. The circle which Mayer invented, and Borda first brought into use in France, is an excellent instrument. Borda's method of calculating the Longitude is very good and expeditious; and he has been as active in introducing the Method of [Lunar] Distances among the French seamen, as Maskelyne was in promoting it among the English. About four-fifths of this report was taken up by the above-mentioned well written historical relation; the remainder consisted of a particular account and opinions of Maingon's memoir. The theoretical principles and algebraic processes, on which Maingon's invention depended, were shortly noticed, and, on the whole, his performance received becoming and well merited commendation. The Institute resolved that the report should be printed; but the form gave rise to debate. Some wished it to be printed separately, and others proposed that it should be inserted in the second volume of the memoirs of the Institute; and this last proposition was carried by a majority of voices. This day's transactions were closed by Prony, who read a letter from Delambre, mentioning that the measurement of a base-line near Perpignan, was very nearly completed.

At the meeting of the 21 vendémiaire, or 12 October, the president gave an account of several French and foreign communications received by the Institute. Dr. Humboldt read a memoir on the application of the principles of modern chemistry to agriculture, and particularly in explaining the effect of manure on the growth of plants. The president read a list of reports of committees, which had not been returned to the Institute, before the current academic year. Some present members of committees promised to bring them in with the required notices: others declared, that the respective authors had withdrawn their plans and memoirs, which plainly indicated a conviction of the impracticability of their proposals, and the inconclusiveness of their deductions. Some memoirs and reports could not be accounted for; and it was conjectured that they had been carried to Egypt by Monge and Berthollet, who had acted as members of committees.

The National Institute had, by circular letters, requested descriptions of the climate, state, agriculture, manufactures, natural productions, &c. of other countries; and when the Institute had nothing particular to attend to, for the two hours of meeting, some of those descriptions were read. It is evident that such productions must have very different degrees of merit. This evening a piece was read, which contained accounts of Greece, Egypt, and Turkey, by Felix, the French consul at Salonica. It seemed to possess no considerable merit, except its describing countries, towards which the national attention was, at that juncture, particularly directed. After it was read, Dessessarts, the physician, who either had not heard, or pretended not to know, the author's name, enquired who had written that memoir. The president answering it was Felix the Consul, Dessessarts excited a general laugh, by rejoining, in his usual facetious manner, "*Felix qui scripsit, infelix qui audivit.*"

Among the transactions of the meeting on the 11 nivôse year VII or 31 December 1798, I shall only mention the very remarkable experiments made on artificial cold by Fourcroy and Vauquelin. These experiments, which were formerly made on a small scale, by Löwitz at St. Petersburg, have not only been repeated, but very considerably extended, at Paris. Within a large tub was placed a smaller one, and the interval between them was filled with a mixture of snow and salt, which produced a remarkable degree of cold. Within the second was placed a third, and the interval between the second and third was filled with a composition consisting of eight parts of muriate of lime,[12] and six parts of snow. In the inner tub was very soon produced an intense degree of cold, which sunk the common thermometer of Réaumur to 32 degrees below zero. In order to keep out the external warm air, the whole apparatus was covered with a glass case. By these interesting experiments, 20 lb. of mercury was made to freeze in thirty seconds into a solid mass, which assumed a crystallized form. Spirits of wine, the strongest vinegar, nitric acid, pure ammonia, and aether, froze in like manner. A finger applied to this mixture or solution, in four seconds lost all sense of feeling, became frozen, and as white as paper, with a very acute sensation, resembling a violent pinch. Most liquors froze, in a platinum crucible, in thirty seconds; but, in a crucible of porcelain or clay, they required about two minutes, which is easily accounted for, from metals being more capable of conducting heat than clay.

The atmospheric cold, when those experiments were performed, was 7 degrees by the centigrade thermometer, or 5.6 degrees of Réaumur's. Decimals being quite fashionable in France, thermometers are used, in which the distance between the freezing and boiling points, is divided into 100 degrees, instead of Réaumur's division into 80 degrees. The Swedes have long used this division, under the name of Celsius' or Christiernin's, thermometer.

On the same evening, Delambre and Méchain related to the Institute an account of their observations on the cold this winter, which has been very severe at Paris, stating, that on 5, 6, and 7 nivôse, the centigrade thermometer stood at 16.2 degrees, and Réaumur's at 13 degrees below zero. The frost first commenced in December 1798; and, except a few days thaw, continued till the beginning of February 1799. For some days in December and January, the thermometer stood at 13 degrees below zero, snow

[12] Calcium chloride.

falling now and then, but seldom exceeding the depth of six inches, and the river Seine was frozen over. Indeed, severe weather is the more sensibly felt in southern countries, because the construction of the houses and the apartments is not calculated to exclude the cold.

I have already observed, that when an author, whether French or foreign, sends any publication to the Institute, the president nominates a member of abilities in the science treated of, to select extracts from the work, and to read them at one of their meetings; a practice which has the advantage of making every member of the Institute acquainted with the contents and merits of the book. These reviews are always well written and impartial, conveying accurate ideas of the contents of every work—not like those critiques in some other countries, which may rather be called reviews of authors and individuals, than of their writings; and which, being composed with a view to introduce the thoughts and opinions of the censors, instead of those of the authors, are more of a didactic than a critical nature.

Many things are sent from the ministers for the opinion of the Institute. Private individuals, in like manner, send in memoirs, drawings, or models of machines, plans of various practical works, &c. in order either to make them known, or to obtain some other advantage. The Class, to which the matters communicated are submitted, always nominates to examine them a committee, whose reports are read and the communications are approved, rejected, or modified. It is natural to suppose that many projects are sent in, which are neither important nor useful; and I have often pitied the members of the Institute, in being obliged to spend much of their time on business of this kind.

Having thus given an account of some of the particular meetings of the National Institute, or more properly of the mathematical and physical Classes, which take place every first and sixth days of the decade, I shall describe their public or solemn meetings. These meetings were not held in the same room as their particular assemblies, but in another much more extensive and beautiful, and which formerly belonged to the Academy of Sciences. Both its longer sides are adorned by two beautiful colonnades; and the ceiling is finely painted and decorated. Between the columns are fourteen beautiful marble statues (seven on each side) of the greatest and most celebrated men whom France has produced; namely, Condé, Tourville, Descartes, Bayard, Sully, Turenne, Daguesseau, Luxembourg, L'Hospital, Bossuet, Duquesne, Catinat, Vauban, and Fénelon. At the ends, are two sitting figures of Pascal and Rollin. In the antichamber, are the statues of Molière, Racine, Corneille, La Fontaine, and Montesquieu. The hall is extremely well lighted, by chandeliers and silver lamps. The floor is covered with a carpet; tables are placed parallel to the four walls of the hall, at which the members of the Institute took their places. There are particular places for the Directory, the ministers of the republic, and foreign ambassadors.

The president of the Institute is seated at the uppermost end of the hall; and in the middle, and rather on one side of him, is a tribune, from which whatever is proposed is received by the president, who does not leave his chair. The place allotted for members is surrounded by a rail, between which and the walls there is round the whole hall a row of benches, where the spectators (among whom were many ladies) took their seats.

The first public quarterly meeting, at which I was present, was on the 15 vendémiaire year VII or 6 October 1798. Jussieu, the president, opened the meeting in a short speech, wherein he signified that, in the first place, an account of the labours of the National Institute for the last three months, would be given by the secretaries of the different Classes. Lassus, after an extemporaneous preamble, read a well written abstract of the labours and memoirs of the physical Class; Lefèvre-Gineau stated those of the mathematical; Daunou those of the moral and political Classes; and lastly, Andrieux read abstracts of memoirs relating to the fine arts. In particular, he gave an account of a dissertation by the famous Dupuis, who wrote the *Origine des Cultes*, and many other well known works, in which the author endeavoured to prove that Denis, the *ci-devant* tutelar saint of France, was no other than Bacchus. As this must be a very acceptable sentiment to every Frenchman who is fond of wine, it was received with a general plaudit.

The president then delivered a short speech, on the progress which the arts must necessarily make among a people, where they are cultivated, esteemed, and rewarded, and then crowned with green wreaths, those pupils who have received prizes in the fine arts.... These industrious young artists, by obtaining the first prizes, have acquired the right of being sent to Rome, whenever circumstances will permit, and there prosecuting their studies at the expense of the republic.

In the next place, Camus delivered an extemporaneous discourse, and gave an account of other great and important labours in which the National Institute were engaged. Under the monarchical government, the Academies of Sciences and Literature had begun different works of importance to mankind, and which on that account would reflect honour on the nation. They intended to publish, 1. The whole of the French historical writings; 2. French and foreign diplomatic papers; 3. A catalogue of the manuscripts in the National Library; 4. Descriptions of arts and manufactures. These designs were interrupted by the Revolution; but every friend to science and literature must hear with pleasure, that these important labours are to be again undertaken, and that the present government will grant the supplies necessary for that purpose.

The National Institute have nominated committees, who are to proceed on the plan of those, who, under the former government, laboured on collections and editions of the old French historical writings, such as Brial and De Clément, the famous author of "L'Art de verifier les Dates." These committees are also to confer with Dutheil and Brequigny, concerning a diplomatic collection. Camus assured the Institute, that a volume of the old historical writings collected by Brial and Druons, and another of diplomatic papers, collected and published by Dutheil, would be sent to the press in about a month.

The National Institute intends publishing a collection of Crusade histories, which are important monuments of the history of the Eastern and Western countries, from the eleventh to the fourteenth century. Hitherto the histories of the Crusades have been related to us only by Western authors. But it is equally important for us to know the accounts of the Orientals, and to see what they thought of the arrival, stay, customs, and victories of the Europeans, with other particulars respecting those invaders. Camus then proceeded to an account of the manuscripts in the National Library, a work which was begun by the Academy of Inscriptions,[13] in the year 1785. Their design was to give moderate abstracts of the less important manuscripts but complete translations of the most valuable, and, in some cases, the manuscripts themselves in their original languages.

The Academy had appointed eight commissioners, of whom three undertook to examine the Oriental manuscripts; two, those in Greek and Latin; and three, those of the Middle Ages. Those commissioners had published four volumes of *Notices des Manuscrits de la Bibliothèque du Roi*. This work, so auspiciously begun, is now carried on with all possible zeal, and the business appears to be of the greater concern, as the number of manuscripts in the National Library is considerably augmented by others brought hither from Italy, Flanders, and Germany, and from the libraries of emigrants, and abolished cloisters. The Institute has particularly in view such manuscripts as concern the sciences, arts, history and geography. The Arabian and Persian manuscripts which relate to astronomy, geography, and history, are to be first published. The Arabians have undoubtedly a number of important and useful astronomical observations, the comparison of which with modern astronomy will be a great acquisition. Nothing is wanting but a good translator, who can comprehend the true meaning; it being a great disadvantage if the orientalist be not an astronomer, or if the astronomer be not a complete orientalist. Camus reported, that considerable progress had been made in the impression of the first volume of the new collection of manuscripts, being the fifth of the whole collection. It contains an account of Oriental, Greek, Latin, and French manuscripts,

[13]"Academy of Sciences"—J. J.

concerning natural and civil history, morals, and the arts; and will afford considerable knowledge respecting the sciences of the twelfth, thirteenth, and fourteenth centuries. Specimens of the original manuscripts, in their respective characters, are to be printed in this volume, which will make it an important acquisition to palaeography.

Camus next proceeded to give an account of the arts and manufactures which the National Institute had cultivated in obedience to an order of government. . . . From the printed programme which is distributed at the public meetings, it appears, that the former Academy of Sciences had either written, or extracted from the writings of others, 87 memoirs on arts and manufactures: whereas those which the National Institute have either caused to be written on the same subjects by its members, or have received from others, amount to no fewer than 297, which are alphabetically arranged. This circumstance is a proof of the industry and attention with which technology has been pursued by the Institute. Among the principal memoirs there are some on subjects altogether new, such, for example, as those on aerostatics, or the method of constructing and managing air balloons; on the art of conducting and maintaining fire; on the art of erecting conductors of lightning; on tachygraphy, or a secret method of writing by signs of abbreviation; and on telegraphy, or the construction of telegraphs, and the signals which accompany them. I might mention various other articles, not immediately reducible to the head of arts and manufactures, such as the projection of maps and charts, surveying and planning, pharmacy, and the method of making anatomical preparations. In conclusion, Camus mentioned the admeasurement of the arc of the meridian, through the whole extent of France, from Barcelona to Dunkirk, and the weights and measures founded on that admeasurement by the commissioners of the Institute, in conjunction with the foreign commissioners, who had come to Paris for that purpose.

Extracts from the memoirs, at the particular meetings, presented to the Institute, during the last six months, were then read by several members. It is natural to suppose, that they selected such pieces as appeared to be the most important and interesting. Cuvier read a description of portions of skeletons found in quarries, in the country about Paris, and particularly at Montmartre. It so happened, that he had collected such a number of bones, as to be able to compose the complete skeleton of an animal. He believed that it formed a new species, which ought to be placed between the rhinoceros and the camel; but this is the only animal, known at present, which belongs to it. Dr. Dessessarts showed that the small-pox, which then generally prevailed, would become less fatal, by preparing the children with jalap and certain mercurials.

Bougainville, the celebrated mathematician and circumnavigator, read an historical detail of ancient and modern voyages towards the north pole. He made a comparison between the situation of sailors in a naval engagement, and on a voyage of discovery. He touched on the voyage of La Pérouse, and the naval engagement on the coast of Egypt, with much elegance and patriotic zeal. The whole of his memoir was exceedingly engaging . . . and well expressed. Bougainville's prelection was often interrupted and finally followed by general, loud and ardent plaudits, which were much more respectful to him than the trifling marks of approbation, often dictated by mere civility, which were given to the other speakers.

The learned Langlès read a memoir on the Arabian language and literature. It is known that the Arabians were men of science and zealous cultivators of the mathematics, particularly of astronomy, when all science was banished from Europe, and their literature is interesting and important. I have to express my concern that Langlès' voice was so low and indistinct, that a great part of his speech could not be understood.

Lacépède read a memoir on the comparative degrees of industry and sagacity observable in birds. He distinguished them into eight classes, according to the sagacity indicated by the construction of their nests, which was the criterion, he adopted, and he named the birds, which he supposed should be referred to each of those classes. Lacépède, with the advantage of an excellent voice, possesses much eloquence, propriety, and dignity, and his memoir was received with general approbation and clapping of hands.

Daunou read a programme, written by Roederer, respecting the question proposed by the Class of moral and political science: "What are the most proper principles on which the morals of a people can be established?" Of sixteen answers given to this question the year before, not one obtained the prize. The same question was repeated, with new conditions and limitations, in order to give the authors an idea of the necessary reply, in which all the former candidates had failed.

The celebrated Fourcroy read an extract of a memoir on the analysis of human *calculi*, together with an account of some experiments made to dissolve them, after being extracted out of the bladder. The memoir was excellent, and admirably delivered.

Bitaubé was to have read an account of the opinions of the philosophers of the ancient republics, but was prevented by want of time, added to his great age and low voice. Ducis delivered a beautiful poem, abounding with enthusiastic encomiums on the fine arts, and the admirable performances of the French painters, Taillasson, Vincent, Regnault, Vien, and David; and with this piece, concluded this truly great and interesting meeting.

I was also present at another general meeting on 15 nivôse year VII or 4 January 1799. Lainée then read an account of the labours of the moral and political Class and Andrieux of the Class of literature, and the fine arts. After mentioning the conquest of Naples, he concluded with expressing a wish that it might not be long before the museums of Portici should be brought to Paris. . . .

Lefèvre-Gineau read a report of the mathematical, and Lassus of the physical, labours of the Class devoted to those pursuits. They also gave an account of the National Institute at Cairo, and of their meetings and transactions, according to the notices which had been communicated to the National Institute at Paris. The transactions of the physical Class were particularly interesting.

L'Héritier, who is acquainted with, and on many accounts highly esteems, the industrious Wiborg, presented a description of two new *genera* of plants, namely, the *Bruguiera*, and the parasitical plant, *Rhizodendrum*. Michaut has seen a tree named the *Robinia viscosa*, from North America, which has on its branches, when in vegetation, a black and strongly glutinous substance. Vauquelin has examined it, and found it altogether different from every vegetable production hitherto known; but it nevertheless approaches nearer to resin than to any other substance. Cels and Ventenat have shewn, that this tree belongs to a *genus*, described by Jussieu and Lamarck. Desfontaines has sent to the Institute a complete Flora of Mount Atlas. Broussonnet, who has long resided in Africa, has particularly described the processes used at Fez and Tetuan, in preparing and dying Turkey leather, and has given an account of the plants employed for that purpose. Lamarck has formed a classification of shells, after a new system and characters. Linnaeus had only sixty *genera*; but Lamarck has extended them to one hundred and seventeen, by which he supposes the classification of shells will be more certain and better determined than formerly. Fourcroy and Vauquelin, by some experiments on urine, have discovered a particular animal substance which gives it the property of very readily forming ammonia; yet they look upon their investigation of the properties of that fluid, as very far from being complete.

The Class of literature and the fine arts proposed for a prize for the eighth year the following subject: "To point out the means of causing the Latin and Greek languages to be cultivated in France, more zealously than they are at present." The prize offered, is a medal of eight hectograms, or about twenty ounces, of gold. The same Class have also proposed a prize, of the same value, for solving this question: "To enquire, to what degree the French language has acquired perspicuity and elegance, and lost its natural simplicity and energy, from the time of Amyot to the present day?"

The mathematical Class have selected the important and very difficult problem of the comet of 1770, which may be considered as an astronomical enigma. The Academy of Sciences, in the year 1794, offered a prize for the calculation of this comet, and the astronomers have attempted to bring their observations to correspond with a parabolic curve. . . . Solutions must be sent in before 15 messidor year VIII or 3 July 1800. The prize is a kilogram, or something more than 2 lb. of gold. But the question is so very difficult, will require so much penetration and labour, and involves such an incredible number of calculations, that, upon the whole, it deserves a greater prize; suppose from six to seven hundred dollars, or from one hundred and fifty to one hundred and seventy-five pounds sterling.

The first volume of the Memoirs of the National Institute was published on 1 vendémiaire year VII or 2 September 1798, and printed by Baudouin, under the title of *Mémoires de l'Institut National des Sciences et Arts. Sciences Mathématiques et Physiques,* 1 tom.—*Sciences Morales et Politiques,* 1 tom.—*Littérature et Beaux Arts,* 1 tom. In all, three quarto volumes, with twenty-four plates; price on common paper, thirty-nine francs, on strong paper, sixty francs, and on vellum paper, seventy-two francs.

It is singular that Baudouin refuses to sell the memoirs of each Class separately; but obliges the purchasers to take all the three volumes. I could not persuade him that he lost, instead of gaining, by this method. The volumes of the mathematical and physical Classes are chiefly confined to natural history, chemistry, and medicine. There are only two mathematical memoirs, one by Laplace and the other by Lalande; for the mathematical members of the National Institute publish their works themselves.[14] Thus Lagrange has lately given the world two important works, namely, his *Théorie des Fonctions Analytiques,* and *Resolution des Equations Numériques*; nor is it long since Laplace published, *Ex-*

[14]Bugge makes an important point here because it indicates a change of emphasis in science imparted by the Revolution. Instead of publishing memoirs intended solely for the elite of the Institute and other professional colleagues, the mathematicians were more concerned to publish books which would reach a wider public, often textbooks for their students. Bugge sees (p. 98) a further reason for the few mathematical papers in the *Mémoires* of the Institute. Some of the newly established institutions of higher education had founded their own journals and were claiming for publication papers by members of their faculty, who were often members of the Institute.

position du Système du Monde. This work appears to be an introduction to his *Traité de Mécanique Céleste*, in two volumes, which contains the discoveries and opinions of this great mathematician, in the theoretical and higher parts of astronomy.

Newton laid the true foundation of our knowledge concerning the order and disposition of our system, and the motion of the planets in their respective orbits. Laplace has finished this beautiful fabric and, with infinite sagacity, has, by help of the higher analysis, in which he is so distinguished a master, clearly proved, that all the motions and phenomena in the planetary system can be explained, determined, and calculated by the principle of universal gravitation, which was not before, in every respect, completely effected.

Laplace is at present engaged on the mechanism of the planetary system, and I have seen about half of the first part already in print. Dr. Burckhardt, of Gotha who studies astronomy under Lalande, translates every sheet, as fast as it is printed, into German; so that the German translation will appear at the same time with the French original.

Bossut, already well known for his mechanics, statics, hydrodynamics, &c. has lately published *Traité de Calcul différentiel, et de Calcul intégral*. . . .

Prony has just published in 4to. *Exposition d'une Méthode pour construire les Equations indétérminées, qui se rapportent aux Sections coniques*. He is besides occupied on a third volume of his very respectable work, *Nouvelle Architecture Hydraulique*, and on the elements of the mechanical sciences. Legendre, who formerly wrote *Élémens de Géométrie; Mémoire sur les Transcendantes elliptiques* and *Dissertation sur une Question de Balistique, couronné par l'Académie de Berlin*, has lately published an excellent work in quarto, entitled, *Essai sur la Théorie des Nombres*.

Lalande is engaged on a complete *Bibliographie Astronomique*. Beside the profound and enlarged views of this gentleman in astronomy and its kindred sciences, he is a great literary character. His extensive reading and correspondence have furnished him with details from every country; so that a complete account of the astronomical writers and literature, of all nations, may be expected from his pen. I have communicated to him all that I could collect on this subject in Denmark.

Messier is continually occupied in discovering comets, and calculating their paths. Delambre and Méchain have measured nine degrees and a half of the meridian of the observatory of Paris, and which stretches quite through France, from Barcelona to Dunkirk. Delambre has been employed on an important work, which he laid before the Commission for Weights and Measures, under the title of *Méthodes analytiques pour la Détermination d'un Arc du Méridien*, and which is now in the press.

Borda, though aged, infirm, and consumptive, still labours as much as his health will permit. He is now engaged on a manual of tables of logarithmic sines, after the new centesimal division. I saw at his house, several printed sheets of those tables; but he complained that, on account of the want of good and uniform paper in France, the impression proceeds but slowly. The sines and tangents are to be found in the stereotypic edition of Callet's tables (Paris, 1795) but not divided into centesimal minutes. Borda's edition will be much more complete, and at the same time more useful. He has also discovered, this winter, some improvements, and new constructions of the barometer and dipping compass. Both these instruments are to be excecuted by that able maker Le Noir. The principal improvement in the dipping needle is, that its axis turns in a glass cylinder or tube. I had formed the same idea many years ago, and have since had a compass so constructed, which I have described in the *Memoirs of the Copenhagen Academy of Sciences*, fourth part of the new series, which contains a drawing of this instrument, and an account of the observations made with it.

As another probable cause of the paucity of mathematical memoirs in the first volume of the Transactions of the National Institute, it may be remarked, that most of the members of this class are lecturers in the Polytechnic and Mining Schools, and other institutions, and that the journals published by those schools contain many of their memoirs, which is the case with Lagrange, Prony, Lefèvre-Gineau, Brisson, Haüy, and others.

The writings of the old Academy of Sciences were divided

into two parts, *Histoire* and *Mémoires*. The first contained an historical account of its proceedings, and extracts from the minutes; and the other, the memoirs themselves. Since the organization of the present National Institute, no part of its history is admitted into its writings; but, in the general meeting at the close of the year, a particular account of its proceedings is delivered by the president of the Institute, to the presidents of the Council of Five Hundred, and the Council of Ancients, who respectively reply to the speech made by the president of the Institute. One of these reports, or accounts of the Institute, is entitled, *Compte rendu et présenté au Corps Legislatif, le premier jour complémentaire l'an IV, par l'Institut National, contenant l'analyse des travaux pendant l'année 4me.* Similar accounts have been published for the years V and VI.

In all that I have said of the members of the National Institute, in this and the preceding letters, my readers will observe that I have always mentioned with warm commendation and becoming respect, those who are eminent in their respective pursuits, and that I consider the National Institute of France, as being one of the most important learned societies in Europe. If I should think or write otherwise, I should look upon myself as destitute of all understanding. Hence I was the more surprised, when, after my return home, I found it signified, in the *Décade Philosophique*[15] that in my letters to Copenhagen, I had uniformly reviled the Institute, held it up to ridicule, and depicted it in the darkest colours.

I need make no observation on the meanness of slanderously misrepresenting the correspondence of an individual with his friends, merely to find, or make, a pretence for complaint; since success in persuading people to believe his aspersions ultimately fixes a stigma on the calumniator himself. But I do hereby deny my having ever written a syllable with which the members of the National Institute, either individually or collectively, could be offended; and I challenge any person whatever to produce a letter from my hand having that tendency. In compliance with the

[15] *Décade Philosophique,* No. 15, 30 pluviôse an VII (18 February 1799), 372.

advice of my friends at Paris, I have made no reply to any of those libellers. My friends know that their assertions are untrue, and those who are not acquainted with me will be convinced of their falsehood by the publication of my travels. On this disagreeable subject I have been hitherto silent; and have looked upon my puny assailants in the *Décade Philosophique*, with that contempt which they deserve.

Chapter 3

OBSERVATORIES

The section dealing with the Paris Observatory is one of particular value, for the judgments of the Danish Astronomer Royal have, of course, special authority. It is perhaps a matter of regret that in 1798 the Paris Observatory was very far from its best. In this chapter Bugge goes into rather more technical details than usual.

Tycho Brahe's observatory of Uraniborg was destroyed soon after he left in 1597, and a hundred years later Paris and Greenwich could have each claimed the title of international center for astronomical observations. The Paris Observatory, designed by Claude Perrault, was completed in 1672. The claim of the French to a predominant place in the history of astronomy and geodesy in the late seventeenth century was based not only on the work of Adrien Auzout, Jean Picard, and Jean Richer, but also on that of the great foreign astronomers who were persuaded to work in Paris: Jean Dominique Cassini (the first of a line of famous Italian-French astronomers), who lived in Paris from 1672 until his death in 1712, and Olaus Rømer (1644–1710), the Danish astronomer who spent six fruitful years in the French capital.

Up to the time of its reorganization in 1784, the Paris Observatory had been dependent on the Académie des sciences. There began in 1786 an extensive program of reequipment to bring the instruments up to date, but it was suspended in 1791 because of the political situation. The director of the Observatory, Cassini IV, had been to England in 1787 to collaborate with the British authorities on an extensive program of triangulation. He had taken advantage of this visit to ask the French government to buy some British instruments, then the envy of the rest of the world. One of these instruments was an eight-foot mural quadrant, and the arrangements for setting it up were still being made at the time of Bugge's visit.

Bugge has some interesting observations to make on Cassini IV (who, like his great-grandfather, had the Christian names Jean Dominique), and had already had some communication with him before the Revolution. Cassini, an ardent royalist, had resisted the moves made by the Convention and the Committee of Public Instruction to impose republican principles on this old establishment. In September 1793 it was decided, in conformity with strict egalitarian ideas, to abolish the established hierarchy in which three pupils of varying seniority were responsible to a director. All were to be equal. Cassini found such a situation intolerable and he resigned. The annual publication of astronomical observations, which he had begun in 1785, ceased and was not resumed until 1801. After a period of mild anarchy, responsibility for the Observatory was taken over by the Bureau des Longitudes, created by decree, 25 June 1795.

Although the Observatory was rather far from the center of Paris, it was visited by a number of tourists because it provided, long before the Eiffel Tower, a high point from which to view the French capital. Friedrich Meyer, who visited it in 1796, made the following comment:

> The view from the flat roof of the Observatory, situated at one of the extremities of Paris, dominating this immense town and its surroundings, is immeasurable in extent and of great beauty. At present this is what is most worth seeing at the Observatory, the interior arrangement of which is still to be carried out. Large vaulted rooms are still being built and in government depots a large provision of astronomical instruments are being

> kept so that a complete collection of instruments can be set up there.[1]

The roof of the Observatory was also used as a telegraph station.

The National Observatory

You will readily believe that the National Observatory appeared to me one of the most interesting places that I had seen. It is situated near the farther end of Rue St. Jacques. The length of this street, from Pont Notre Dame to the Barrier, is 1275 *toises*. The observatory stands 150 *toises* from the Barrier, on an eminence, and, like the whole of Paris, on a chalky basis. This outlet of the city not being much built upon, there is much open space about the observatory. It is not incommoded by smoke and damp, and possesses a free air, and a fine prospect. We Danes claim the honour of having been the first nation in Europe who . . . erected solid and durable observatories. Who has not heard of the immortal Tycho Brahe's Uraniburg at Hven? Who does not know that, after this great man's exile from Denmark, Christian VII, without doubt, lamenting this loss to the sciences, caused the round tower at Copenhagen to be built, and there fitted up an observatory for Christian Longomontanus, the most famous disciple of Tycho? The observatory at Copenhagen was finished in 1637: and it was not till thirty years after, that the observatories of Paris and Greenwich were built, almost at the same time. The establishment of the Academy of Sciences, and of the observatory at Paris, owe their origin to the anxiety which the great Colbert, Minister to Louis XIV, felt for the promotion of sciences. The observatory was erected by the celebrated French artist Perrault, who has paid more attention to the beauty of the edifice, and to his own fancy as an architect, than to the accommodation of astronomers. The building consists of two very large and high stories; all the floors are in good order, and on the roof is a

[1]Meyer, *Fragments sur Paris*, II, 76.

platform or gallery. Under the building are cellars of remarkable depth, and which I shall hereafter particularly notice. The finest front is the least seen, as it faces a garden belonging to one of the residing astronomers, who at present is Méchain; so that from the common entrance in Rue St. Jacques, the observatory appears to some disadvantage. This establishment was falling into decay during the latter years of the monarchy; at least some of the instruments were so old, that others, suitable to the present improved state of astronomy and mechanics, had become absolutely necessary. Count Cassini, who was at that time Director of the Observatory, represented to the Government the deficiencies complained of, and had actually begun to make them good. But the Revolution took place, Cassini was obliged to quit the observatory, and the building and instruments were greatly injured in the times of terrorism. When that direful period of frenzy was past, and the arts and sciences were again thought of, astronomy and the National Observatory were not forgotten. It is now undergoing a thorough repair, which is much wanted; and it is to be supplied with instruments corresponding to the present perfection of science.

When I first visited the observatory, I found below, in a kind of roomy and well furnished cellar, a door open, and an old man sitting at a table. Supposing him to be the porter, I enquired for Méchain, Delambre, and Bouvard. He told me, that Méchain and Delambre were gone to Perpignan, in order to measure a base-line for determining a degree of the meridian. The supposed porter had papers before him, containing geometrical figures and algebraic calculations. I asked him, if he amused himself with geometry and algebra? "Yes, in part," replied the venerable man, "but chiefly with astronomy. I was formerly astronomer of the observatory, but am now, as you see, thrust down into this cellar." "Your name?" "Jeaurat." "And I am Bugge, from Copenhagen, who highly esteem you, and am well acquainted with your former labours." It gave me great pleasure to become acquainted with this worthy man, who calculated the *Connaissance des Temps* from 1776 to 1787. Jeaurat, who is the oldest of all the present astronomers of the Parisian Observatory, established and put in order a similar erection at the military school, and is the author of thirty essays in the Memoirs of the Academy. It happened to him, as to many more, during the Revolution, to be supplanted by younger rivals of superior interest, though not always better qualified. This astronomer, in his seventy-second year, has nothing to live upon but the salary of the youngest member of the National Institute, which is 1,200 francs, two small apartments on the ground floor, and a little garden. I requested him to have the goodness to shew me the observatory; but he declined it, and deprived me of an opportunity of thanking him. I was then obliged to enquire who superintended the observatory in the absence of Méchain and Delambre? And was answered, Bouvard, adjunct astronomer, who lives in a small separate building belonging to the observatory, and where Méchain formerly resided for twenty years. Bouvard again unluckily was not at home, and I was obliged to content myself with the Citoyen Portier, a follower of St. Crispin, who, for the last eighteen months, had made shoes, waited at the observatory, and shewed it to strangers, and I had great reason to be satisfied with his service.

On the first floor, apartments are being fitted up for Méchain, who lives at present in those which Cassini formerly occupied; and on the other side of the principal passage, Messier is to be accommodated in rooms, which are now under repair. There is, on the same floor, a spacious apartment for the use of the observatory, from which is an entrance to the side building, where transit instruments, and mural quadrants are set up, of which I shall give a more particular account. The whole observatory being at this time under repair, the instruments have been laid aside wherever convenience allowed. The following are in a lower apartment: 1. A brass equatorial instrument, made by Haupois, in 1792, for measuring the declination, having affixed on each side a circle of two and a half feet in diameter. It must be acknowledged that Haupois is a good workman, as the divisions appeared accurate, and the whole well finished: but the instrument itself is by far too complex and troublesome to be used in observations. 2. A brass quadrant, by Haupois, 1793, of eighteen inches radius, very well made: this is generally used for taking corresponding altitudes. The stand appeared to me to be very weak, and not sufficiently steady. 3. An excellent astronomical timepiece, by Berthoud, with a pendulum to correct the errors arising from the influx of heat, which is in fact nothing more than Harrison's gridiron pendulum. 4. A reflecting telescope of five feet, by Dollond. The stand is exceedingly steady and strong; this telescope is set up like an equatorial instrument, in order to assist in observations taken off the meridian. It is a thoroughly good instrument, executed in such manner as might be expected from an English artist of Dollond's abilities. I requested leave to look through this telescope, which the porter granted me, with greater readiness than I expected. I tried it on a very remote object, and found it exceedingly good. In this apartment are the busts of Colbert, Jacques Cassini, Dominique Cassini, and Maraldi, all of gypsum.

In a smaller apartment adjoining the large one above described, was a three foot quadrant, made by Langlois in the old French manner. This quadrant has over it a kind of moveable cap, or shade, and has been long used in taking corresponding altitudes. The French of late have begun to use the meridian circles, called by them *Instrument de Passages*, and which was invented by our countryman Rømer, under the name of *Rota Meridiana*. This instrument has been used by the English ever since 1716, and termed by them a transit instrument. La Caille formed his catalogue of the fixed stars, by corresponding altitudes, and without the help of that instrument.

I next went into a room, intended for transit instruments, and mural quadrants, and which was under repair, the floor and ceiling not having been finished. We then went up into another floor, which is very lofty: the height, I should suppose, might exceed twenty Danish ells.[2] In the middle is a large room, on the floor of which is drawn a meridian line. It should be observed that meridian lines were, at one time, in great vogue in France.

[2]Twenty Danish ells is approximately forty feet (1 Scandinavian ell = 2 local feet; 3.1885 Danish feet = 1 meter).

I have found them drawn on floors at Versailles, St. Cloud, Trianon, Chantilly, and many other places. There is nothing in this room but a large telescope of sixteen feet focus, which formerly belonged to the King, and had been set up at Passy. Its metallic specula have, by negligent treatment, lost their polish, and are totally spoiled. At the bottom of the stand, is a large cog wheel, acted on by a pinion, by moving which the telescope can be shifted horizontally from one direction to another. From the centre of this wheel rises a strong axis, to support the telescope, to which is fixed a semicircle, which, by means of a pinion, raises or lowers the instrument; so that by this horizontal and vertical motion, the heavens can be swept by various altitudes. The tube of the telescope is metal painted blue with oil colours. Upon the whole, this instrument has but a mean appearance, and is of no value. Indeed it ought to be laid aside, unless the specula be newly polished.

On one side of this large room are two apartments, one of which is appointed to be the library of the observatory, and the other to be the residence of Delambre. On the opposite side is an apartment with windows in three directions, which contains [several] instruments. . . .

On the other floor, which the singular ideas of Perrault, the architect, intended for the sole purpose of an observatory, neither transit instruments, mural quadrants, nor sextants, can be set up; so that the large and important astronomical instruments, which are most wanted in an observatory, are here totally useless. The floor underneath it is of no service, except it be by fixing a telescope in one of the windows, in order to observe the eclipses of the sun, moon, and satellites of Jupiter, or the occultation of fixed stars by the moon; and then there is an inconvenience in taking down the time by some other chronometer, or by signals, according to which the moment of observation is to be determined; and the calculations for determining such times are always subject to great uncertainties.

Whoever has seen, or been otherwise made acquainted with, the observatories of Greenwich, Oxford, Edinburgh, Mannheim, Gotha, and Copenhagen, will find that their arrangements, though much more simple, are also far more complete and commodious, for all kinds of astronomical observations, than this of Paris.

The porter, in the last place, conducted me up to the platform, where, on a fine clear day, one has an excellent view of this great city, so very remarkable, not only for its scientific, but its political history.

When I had satisfied myself with viewing as much of the observatory as the porter could shew me, I gave him my card to deliver to Bouvard. The porter reading my name, cried out, "Ah! I know you very well—I am much surprised at it—this great honour I could not have expected—stay a little, and I will convince you that I am right." He went into the library, and brought out different numbers of the *Connaissance des Temps*, in which he shewed me several astronomical communications of mine to Lalande and Méchain. "You see now," said he, "that I am not mistaken." I was rather struck with this fantastical shoemaker, this door-keeper of the heavens, who seemed to be so familiarly acquainted with me, merely from his having read my name in the astronomical calendar. A few francs having satisfied him for his attention and trouble, we parted the best of friends; and he bawled after me, "Come again soon, I am always at your service!"

It was not long till I revisited the observatory, though the nearest distance to it, from the place where I reside, in Rue [St.] Honoré, is half a Danish mile. After I had paid my respects to the father of the Parisian astronomers, the aged Jeaurat, in his cellar, I enquired for Bouvard, whom I found at home. With all possible goodness and complaisance, he shewed me the observatory, and such curiosities as were inaccessible to my astronomical shoemaker. Bouvard had in his apartment, 1. A silver watch, or chronometer, made by Berthoud, and which belonged to Borda. This had been on trial for fifteen months, and was found to keep time well: it cost one hundred Louis-d'or. I wished to see whether it was of the same construction as Arnold's chronometer, of which I have from twelve to fourteen upon trial at Copenhagen, and have found them all keep time excellently; but Bouvard could not inform me, as he was unacquainted with its internal structure. I then desired to have the piece opened, that I might see the construction of it. But this again could not be done, as there was a cap screwed down over the work. 2. Coulomb's declination compass: to prevent friction on the supporting pin, the needle, in this contrivance, is suspended by a silk filament, as spun by the worm. The idea is altogether excellent, but still it is not easy to make the centre of the circle described by the needle steadily coincide with the centre of the graduated circumference.

Besides what I had before seen, and now re-examined, at the observatory, Bouvard shewed me, 1. The platinum specula of Rochon's telescope. The great speculum was very good, yet there was here and there some dark speckles on the surface of it, which were undoubtedly owing to the platinum not being perfectly purified before the speculum was cast; and it still contained some small quantity of iron. The little speculum, however, was particularly excellent and clear, and of a beautiful polish so that there are no doubts left, that excellent specula for reflecting telescopes can be made of platinum.

The frame which fastens and supports the speculum was of iron, forming a square of two inches. In the corners of which were screws fastened, which, by pressing on the back of the speculum, kept it from falling by its own weight. But this must be done with great care; for the figure of the speculum would be changed by much pressure. Bouvard has now invented another contrivance for fixing the great speculum, namely, by inclining it a little; so that the object is thrown to the other side of the farther part of the tube. He then took the least or farthest speculum away, and disposed the image so as to be in a right line with the eye-glass. Herschel has very properly made use of this method, in his large telescope of above twenty feet; and it can always be used to advantage in telescopes of a smaller size, in which the head of the observer intercepts too much light, that the image must necessarily be indistinct. 2. Bird's mural quadrant of eight feet radius, which formerly belonged to Le Monnier, and which will be an excellent instrument when set up on its wall, and properly adjusted. 3. De la Hire's mural quadrant of five feet radius, all of iron. 4. The elder Cassini's mural quadrant of the same metal, and of six feet radius, with a brass adjuster. It is divided by dots into spaces of five minutes, and the divisions are taken by a micrometer, after the old method of the French astronomers. The two last instruments are not to be set up again, but are to be looked upon as venerable agents in the service of astronomy; since it was with them that so many observations were made in the early part of the present century. 5. Among the astronomical antiquities, are also to be seen eight or ten object glasses, of eight, ten, and twelve inches in diameter, and of sixty, eighty, one hundred, and one hundred and twenty feet focus, by the Italian, Campani, who, at the close of the last century, was as famous for his refracting telescopes, as Herschel is at the present day for his large reflectors. I must here observe, that the optical and astronomical rhodomontade of a gigantic reflecting telescope of sixty feet, with a platinum speculum, said to have been made here, has no foundation, and has not been heard of, except in a German Gazette, and some other newspapers. With the telescope of Campani, the Cassinis, father and son, made many discoveries, such as the satellites of Jupiter and Saturn, the marculae of Venus, Mars, and Jupiter, and their rotations round their own axes, &c. 6. The rough draught of the observations from the first establishment of the observatory till the Revolution. A chasm for some years succeeds; but now everything is re-established in good order, and the arrangement will be still more complete, when the repairs are finished.

The observatory is now under the direction of the Board of Longitude, by whose order Bouvard made me a curious present, being a copy of the large chart of the moon, twenty inches in diameter, which Jacques and Dominique Cassini caused to be engraved, after a series of observations for nine years, viz. from 1671 to 1680.

The plate was lost for about eight or ten years, but has been fortunately recovered and made public. In the National Observatory has also been found a port-folio, consisting of sixty leaves, with original drawings by Le Clerc, one of the best draughtsmen of his age, of singular and distinguished spots in the moon. Observations corresponding to those drawings, and on which they were founded, are all of them in Cassini's own hand-writing.

The present Cassini, about the year 1788, reduced this large lunar chart to a diameter of eight inches, and had impressions of it taken in blue. I am possessed of one of those impressions, which he sent me to Copenhagen as a present. Both these lunar charts of Cassini are much better resemblances of the moon than that of Tobias Mayer, so much valued in Germany. The only thing I have to observe respecting this chart of Cassini, is, that the ridge of mountains, proceeding in faint streaks from the lunar macula, called Tycho, are not distinct or well defined . . . .

I have before observed, that there is no mural quadrant erected in the building appropriated for observations; nor indeed does it contain any instrument of this kind, which I can much commend. The observatory is at this time receiving great improvements, and will be put into a proper state of repair. For this purpose, it has been found requisite to build a solid wall to the same height as the lower floor of the observatory, and to erect a side building on this basis, for the reception of the instruments above mentioned. This last, though it appears to be only a protuberance on the great body of the edifice, is to be the proper and real observatory.

In this side building are three apartments:

1. One for an observer, in which there is a fire-place.
2. An apartment for a transit instrument.

Though this instrument is not yet brought to the observatory, I shall offer a description of it, having been often permitted to see it by that skilful instrument maker, Lenoir. The achromatic tube is five feet in length, and has a very large aperture. The two movements, one for raising and depressing the axis, and the other for bringing the instrument to coincide with the meridian are very good. The extremities of the axis are of bell-metal, and set in triangular plates. A counterpoise balances a part of the weight of the tube. The level is excellent, and can be very conveniently fixed up, and adjusted. These are entirely new improvements. I have an instrument at the observatory of Copenhagen, in constructing which, the same general principles have been observed. . . . In the telescope are five vertical threads, and on this account the eye glass is made moveable, and can be fixed before any of the threads, so as to prevent the line of vision from being bent or indirect. In the year 1777, I saw a transit instrument at Greenwich, adjusted in the same manner. The only thing which appeared to me to claim the merit of novelty, in the Parisian instrument, was the manner of illumination. . . .

The fine particles of matter, floating in the air, are more numerous than is generally supposed by those, who have not had an opportunity of making observations on this subject. These particles will render the thread thick and uneven, and the glass dull; so that annually, or biennially, the instrument must be taken to pieces, and the glass and thread cleaned. The inconvenience of having the instrument to set up again and adjust, must be a great obstruction, in conducting a series of observations, which require an instrument to be in a perfectly invariable condition. But, though these defects are not to be denied, this transit instrument is well executed, and will be found a very fine one, when set up on its proper pillars, which Bouvard told me, are to be of granite.

3d. The third apartment in the side-building is intended for a mural quadrant. In the middle of this apartment, a wall is now building, of the common calcareous stone, generally used at Paris. When this wall is completed, a mural quadrant of eight feet radius, by Bird, will be suspended on its left side. The limb of this quadrant has a two fold division, namely, into ninety and ninety-six degrees; and is constructed in the manner described by Bird, in his *Method of constructing mural quadrants* (London, 1768). . . .

Thus is this little building (the most important part of this colossal National Observatory) provided for the use of some of the most able and eminent astronomers of Europe, Messier, Delambre, and Méchain. Bouvard, the adjunct, who last year discovered a comet, will most certainly contribute his share towards making a proper use of those instruments. Méchain and Bouvard, who alone live at the observatory, make the observations, and record them in very exact and well arranged protocols. Messier and Delambre told me, that not being inclined to change their abodes, they have each of them a small observatory at their own houses. All the four gentlemen are as kind and obliging as they are eminent for their observations and mathematical abilities.

It gave me great pleasure to become personally acquainted with Méchain, after his return from measuring degrees of the meridian, in executing which, he instituted a series of triangles from Barcelona to Rodez, and, with unwearied industry, ascertained the height of the pole at different places, situated nearly on the meridian of Paris. Ever since 1781, I have kept up a regular astronomical correspondence with him, Lalande, and Cassini, who, before the Revolution, was Comte de Thury, and director of the observatory.

I observed that there was one instrument wanting at the National Observatory, namely, an astronomical sector of ten or twelve feet radius. Lalande gave me to understand, that there is at Paris an excellent instrument of this kind, being a twelve foot sector by Graham, used by Maupertuis in measuring a degree of the meridian, and that this famous instrument will be brought to the observatory.

On 10 brumaire year VI, or 31 October 1798, I was at the observatory, in company with Professor van Swinden, and Aeneae, director of navigation, both from Holland, and Professor Trallès from Switzerland, who, like myself, were foreign Commissioners for weights and measures. Cassini met us there by appointment, in order to shew us the instruments he had constructed, and the methods he used for ascertaining the variation of the compass. Before the principal door of the observatory, on a terrace, at the end of Méchain's garden, the instruments of Coulomb and Cassini were erected on a round pedestal of stone, on which a horizontal meridian line was drawn, a vertical section having been also raised on the whole height of the building.

Cassini's instrument is a circle of ten inches in diameter, furnished with a needle of the same length, suspended by a silk filament, after the method of Coulomb, and a Nonius[3] at each end points out single minutes. Over the centre is fixed a vertical stand to receive a small transit instrument with its level, the line of vision of its telescope being made to correspond with the diameter of the instrument at zero.

The principal diameter of the instrument can be set to the meridian by the telescope, and the above mentioned vertical line, on the wall of the observatory, and by a mark on a wall on the other side of the instrument; and its superficies can be fixed horizontally by two screws below the circle. The angle of variation can be thus found, either directly or by doubling it, as with the circle of Borda, in order to obtain the minutes still more accurately. With this instrument of Cassini, the variation was observed on 31 October 1798, by the following gentlemen:

By Bouvard	22° 13′
Trallès	22° 11′
Van Swinden	22° 11′
Bugge	22° 11′
Mean variation of the needle	22° 11′45″

C. Coulomb's instrument has a needle of twenty-four inches in length, and three-fourths of an inch in breadth, suspended by a small wire. It is not a perfect circle, but has at both ends an arc of about thirty degrees, divided by tangents, and over each of

[3]Nonius scale: so-called after the sixteenth-century Portuguese mathematician, Pedro Nuñez or Nonius. In his *De Crepusculis* (Lisbon, 1542), Nuñez suggested a method of accurate measurement of angles by describing within a quadrant forty-five concentric circles. The outermost circle was divided into eighty-nine, the next into eighty-eight, and so forth, until the innermost was divided into forty-six parts. The plumb line or index was bound to coincide almost exactly with one of the circles near a point of division. By calculation the precise angle could be ascertained. This device was difficult to make and was little

the arcs is a microscope. It was unanimously agreed, that there was some eccentricity, and that the needle was accurately suspended. Bouvard, with this instrument, observed the variation to be 22° 12′; so that with both instruments the observations agree very well. Lastly, with a variation instrument, of a construction similar to that of mine at Copenhagen, and which the meteorological society at Mannheim had given to the celebrated Le Cotte, the variation was observed to be 22° 24′, and the excess of twelve minutes might very well be accounted for from an error of the sidereal parallelism with the meridional line, which is properly owing to the sudden friction of the agate on the steel. The justness of the well known preference of Coulomb's method of suspension to the common one, appeared on this occasion very evident.

Cassini went down with us to the cellars of the observatory, which are very remarkable. The descent is by one hundred steps, to the depth of forty feet beneath the surface of the earth. The cellars particularly consist of several labyrinth passages of four feet in width, and five or six in height. In most places those subterraneous passages are walled; but in several the natural stone or rock forms the ceiling, in some places the sides, and in others the floor. These cellars are in general very dry, but in some places, either the ceiling or the floor are moist. In several parts of the ceiling, drops are crystallized into stone and stalactites, and the moisture on the floor is covered with a stony scum or membrane. . . .

When Cassini was director of the observatory, he caused two apartments to be constructed, and separated from the labyrinth by a wall: one of these apartments was designed for observing

used. An improvement was proposed in 1631 by Pierre Vernier of Burgundy. Bugge, following German practice, calls a *Nonius* something which elsewhere was generally termed a *Vernier*. An example of a Vernier Scale is a subsidiary scale on which ten divisions are equal to eleven divisions on the main scale. Each division of the subsidiary scale is, therefore, one tenth greater than the unit measured and it follows that the device can be used to read off directly fractions of one tenth of the main scale. This scale obviously made a major contribution to precision in measurement.

the variation of the compass under ground. In the years 1783 and 1784, Cassini found no sensible difference between the variation above and under ground. . . .*

In the other apartment was a Réaumur's thermometer, made by Mossy, under the direction of Lavoisier.

Every degree of this thermometer was four inches three lines. Cassini made observations by it for three years, and found that the temperature of the earth, or heat of the air under ground, did not undergo a greater change than three tenths of a degree.

These labyrinth caves and large passages under ground, lead to a grate or iron-doors from which there was in ancient times a communication with the quarries: but no man knows how far, or in what precise direction, this passage extends. This grate was set up when the observatory was first built. In taking notice of it, Cassini related to us some of his history in the time of the Revolution, when his going regularly every day down to these caves, in order to observe the magnetic needle and thermometer, gave rise to a rumour among the then ruling Jacobins and *sans-culottes*, and which, as usual, acquired in its propagation, considerable alterations and additions. It was, in short, concluded, that provisions, arms, ammunition, and aristocrats were concealed in the cellars of the observatory. One morning Cassini was very early taken out of his bed, by three or four hundred Jacobins and *sans-culottes*, armed with firelocks, swords, pikes, and cudgels, and forced half-naked to conduct them down to the caves of the observatory, in order to examine those subterraneous recesses. Cassini told them that he obeyed them the more willingly, as he was certain the caves contained none of those articles which they expected to find in them; yet he must tell them before-hand, that the caves of the observatory led to a fastened iron-door or grate, which opened into a hitherto unexplored subterraneous passage, which, for aught he knew, might communicate with places in the city; that he was totally unacquainted with those passages, and of course could not be

*See: Cassini, *De la declination et des variations de l'aiguille aimantée*, Paris, 1791, p. 24.

answerable for what might be found in them. Not half dressed, and surrounded with bayonets, swords and pikes, he was obliged to conduct them through all the caves, and the inextricable windings and meanders of those caves; and this brave band found them, as Cassini had predicted, totally empty. They finally approached the iron-door, which they found had been forced open, probably by some masons and smiths belonging to the troop, while the rest went in quest of Cassini. They demanded that he should conduct them down into the subterraneous passages in the rock: but he reminded them of what he had before said; adding, that he was perfectly in their power, but that he had rather suffer death on the spot, than conduct them down into those unknown passages, for which he neither would nor could be answerable, and that he coolly waited for their decision, even if his death should ensue. The most important among this corps then held a council of war, the result of which was, that Cassini, guarded by six men armed with pikes, should return to his apartments, and that the rest should go down into the passage or cavern. After they had proceeded a good way in, and found nothing, they became tired, returned but again, and spared the observatory for the time. But the edifice has since been often searched, and the instruments, astronomical constructions, and apartments of the astronomers very much injured by such visitations.

Bouvard, though a staunch and zealous republican, told me, that those vandals once took into their heads to sell the observatory, and actually wrote, in large characters, over the door,

PROPRIÉTÉ NATIONALE À VENDRE*

The Cassini, whom I have so often mentioned, began, in 1784, to improve the observatory, to procure new and superior instruments, and to conduct the observations on a better and more accurate plan. He published yearly, from 1785 till 1791, a number or volume of his astronomical observations, on the

*National property for sale.

fixed stars, sun, moon, and planets, calculated and compared with the best astronomical tables, in order to ascertain and correct the errors of those tables. He sent those numbers annually to other astronomers, and he had the goodness not to forget me. He did everything, in short, that could be reasonably expected from an able, industrious, and experienced astronomer.

In the midst of Cassini's celebrated career, the Revolution took place. Having been suspected by the terrorists, he was driven from the observatory, which he had so honourably conducted, and not only deprived of his office and income, but confined in prison above a year; and he has saved nothing but his life, and a small property, which he inherited from his ancestors, where this worthy man, with his numerous family, exists upon a scanty income. In the opinion of some people, the ambition, envy, and egotism of certain other astronomers, have greatly contributed to drive both Cassini and Jeaurat from the observatory.

Among other contrivances, Cassini had a foundry built for casting the large astronomical instruments, which he intended to have had constructed for the observatory. This foundry, when France was filled with factories of salt-petre, powder, and fire-arms, was converted into a cannon foundry. As relics of that direful and alarming period, eight cannons, twelve pounders, still remain there. But the times are so much changed for the better, that the votaries of the beautiful and pacific Urania have now nothing to fear from these dispensers of the thunder of Mars; especially as they are not charged, or so much as furnished with touch-holes.

The best French optician is the able Carochez, of whom I shall on another occasion give a fuller account. Carochez is the only man who has cast the ground specula of platinum, which he did for what is called the Rochon telescope, of six feet. He had the goodness to go with me to the observatory, and to shew me the effect of this telescope, and of his newly ground six foot achromatics, belonging to Borda. The object glass consists of a crown and a flint glass, between which is a mastic effusion (*Mastic en larmes*). The aperture of these conglutinated object glasses, called by the French *collés*, is somewhat more than five inches.

The reflector and achromatic were successively directed towards a piece of paper fixed at the distance of four or five hundred *toises*. This could be very evidently distinguished by both the instruments; but with this difference, that the reflector not only magnified much less, but gave a remarkably brown reflection, and an obscure and confused image: and the telescopes made with the general composition are found to cast a reflection more or less yellow. Carochez's achromatic not only magnifies much more, but has at the same time very great clearness. The paper has its true, and perfectly white colour.

On the last day of January 1799, I paid my final visit to the observatory, partly with a view to take my leave of Méchain, Jeaurat, and Bouvard, and partly to see how far they had proceeded with the apartments intended for the transit instrument and mural quadrant. The apartment for the former was completed, except that the pillars, or pyramidic frusta, designed to support the axis of the instrument, were not finished. . . . The apartment for the mural quadrant was also ready, and the murus, or wall, was built. . . .

I have already said, that the astronomers of the National Observatory are Messier, Delambre, Méchain, and Bouvard; that Messier and Delambre do not reside at the observatory, and that Méchain is an able, industrious, and excellent man, and has laboured in the observatory for thirty years. In the years 1786 and 1787, he measured part of the longitudinal arc, in order to unite the observations of Greenwich and Paris.* From 1792 to 1798 Méchain has been occupied in measuring a meridional arc, from Barcelona to Rodez. He has completed this measurement, and the concomitant observations on the height of the pole, with all the accuracy to be expected from so able a man. Besides the place he holds at the observatory, Méchain is Hydrographic Astronomer to the *Dépôt de la Marine*, or the Marine Depository; and has collected and calculated, with incredible industry, such observations as contribute to determine the

*See: Cassini, Méchain, and Le Gendre, *Exposition des opérations faites en France pour la Jonction des Observatoires de Paris et de Greenwich*, Paris, 1796.

extent of harbours and sea-coasts. From a correspondence of twenty years, which he had held with me, I can bear testimony to the industry he has applied to the northern coasts. Méchain possesses, and deserves the regard and confidence of the Marine Minister, Bruix, and of Vice-Admiral Rossilly, who preside over the Depository. This deserving man, however, is not without his opponents, and I have heard some persons take great pleasure in reviling his character.

Bouvard, who came to the observatory since the Revolution, is a very industrious and skilful astronomer. He always calculates his observations, and compares them with the best astronomical tables. The observations of the Arabians were collected by Ibn Junis, whose principal manuscript is in the library at Leyden. Joseph de l'Isle, who formed an extensive collection of astronomical manuscripts, has a copy of it, which Messier received from him. This is now translated, and Bouvard has calculated out of it, and compared with our modern tables, twenty-six eclipses of the sun and moon, observed by the Arabians from the year 829 to 1004; twelve solstices; and several occultations of the planets; and one of Regulus, or α in Leo, by the intervention of the moon. He has found that the moon must have decreased eight minutes, and that the place of her node has decreased two or three minutes. He has also found, that the Arabians were well acquainted with the mensuration and true length of the year, to within about five seconds of the truth.

The National Observatory is now in good condition, and provided with excellent instruments, for the common use of the four established astronomers, Messier, Delambre, Méchain, and Bouvard. In the present organization of the observatory, no one of the astronomers is subject to another; but all the four are under the control of the Board of Longitude. It is possible that this organization may have its advantages; but it may also be attended with inconveniences. As long as perfect harmony and a good understanding subsist among the astronomers, everything will go on well. But unanimity is not always to be found in this sublunary world of ours, and the interruption of it might be attended with several difficulties. For instance, A undertakes a series of observations, which require that the instrument should neither be moved nor altered. B observes with the same instruments, and unacquainted, perhaps, with the designs of A, he finds that the transit-instrument, or the mural-quadrant, requires an adjustment only of a few seconds. He alters the instrument accordingly, and thus renders A's course of observations useless. Would not confusion ensue, if all the four co-ordinate astronomers should wish to observe at once, one and the same phenomenon; as, for instance, a planet's opposition to the sun, entrance into its node, its aphelion, the inclination of its orbit, &c.? Would it not be better, therefore, to have this observatory organized like all the other establishments of the kind in Europe, where there is one principal astronomer, and the rest assist, and labour under his direction? But if each astronomer had the command of a separate set of instruments, the present regulations of the French National Observatory, would undoubtedly deserve the preference.

On the platform are two small chambers for the accommodation of those who attend the telegraph there erected. In summer, several trials were made there, with flag signals; but I am not acquainted with the object, or the success, of those trials; nor could the astronomers at the observatory give me any information concerning them. I should suppose that there can be no better construction of telegraphs than that with one principal pole, and two arms moveable at the ends; and such is the construction of the telegraph at the house of the Marine Minister, on the corner of the Place de la Révolution, and Rue St. Florentin; of that at the Garde-meuble, which conveys signals to Brest; of that on the Louvre, to Lille; and of that on the church of St. Sulpice, to Strasbourg; and these are the only telegraphs at this time in Paris.

I now proceed to the inferior observatories at Paris, which are to be considered either as public or private; the public are those of the Military School and French College, and the private ones are at the houses of Messier and Delambre.

The observatory at the Military School was put in order by Jeaurat, and Lalande has since had it under his direction. It is on the third floor of one of the wings of the Military School. The walls whereon the observatory and instruments rest are entirely solid, all the way up, and carefully overlaid; so that nothing has been neglected to ensure every possible degree of steadiness. The instruments are, 1. An excellent eight-foot quadrant, by Bird. 2. A very fine four-foot transit instrument, by Lenoir, constructed entirely like the meridian telescope, at the National Observatory, which has been already described. . . . Here are also a time-piece, a smaller quadrant, and several telescopes. Lalande has several apartments at this observatory, though he resides at the *Collège de France*. At this observatory, which is very well constructed, that astronomer and his nephew observe the many thousand stars, which are most of them telescopic, or invisible to the naked eye.

The observatory at the *Collège de France*[4] is on the third floor, and contains, 1. A small transit instrument of three feet, which is not well fixed. 2. A common French quadrant of three feet radius, with a moveable shade over it, intended for taking correspondent altitudes. 3. A four-foot sector. 4. Several time-pieces and telescopes. 5. Borda's circle of eighteen inches in diameter, by Lenoir, which, in my opinion is the best instrument in this observatory. There are also several other less important and older instruments.

The observatory at the *Collège de France*, as well as that at the Military School, is under the inspection of Lalande. This learned man is the Ptolemy of our age; for, as the *Almagest* of that old author contains a complete body of ancient astronomical knowledge; so Lalande's astronomy is a complete and excellent depository of the modern improvements in that science. He is a man of very extensive reading, is well acquainted with all the astronomical writers, and possesses great literary knowledge, qualifications not altogether common, even in France. He has greatly improved the astronomical tables, by determining and calculating their first principles, from the best modern observations. In conjunction with his pupil, and intimate friend,

[4]"French College"—J. J.

Delambre, he has calculated new tables of the planets, which are inserted in the latest edition of his astronomy, and are the best tables of the kind extant. Lalande has a very extensive correspondence with all the astronomers in Europe, who have laboured to promote his favourite science. But it is admitted that he has lost some of his reputation in Paris, and that sufficient justice is not done to his merit.

Messier, so celebrated for the many comets he has discovered, lives at Rue des Mathurins, Maison de Clugny, No. 334, and has, on the upper floor, a small observatory, containing an accurate meridian line, a time-piece, a quadrant, and a parallactic machine, wherewith he has discovered and traced his comets. This worthy old man is very lively and cheerful, as I have experienced in the many agreeable hours I have passed in his company. Apartments have been prepared for him at the National Observatory; but his age, and the convenience he finds in his present situation, and in the use of his own instruments, sufficiently account for his continuing in his old abode.

Delambre, one of the best and greatest astronomers of France, lives in Rue de Paradis au Marais, No. 1, where he has a neat little observatory, with a small transit instrument, a good time-piece, which formerly belonged to La Caille, the two circles, which Borda used in measuring a degree of the meridian in France, and several good telescopes. From a series of more than eight hundred observations, on different circumpolar stars, with the largest of Borda's circles, he has this winter determined the polar altitude, at his observatory, to within a second, or at least as nearly as the truth can be approached with instruments of fifteen or eighteen inches in diameter. Delambre told me, that, when he transferred the latitudes of his own observatory to the National Observatory, allowing for the known difference in latitude, ascertained by the measure of the meridian, he found the latitude of the National Observatory to be what Lalande had made it. I must, however, observe, that owing to uncertainty of refraction, even when Bradley's table is used, as being the most accurate, an uncertainty of a second in latitude may still remain.

Experience has proved, that the Board of Longitude in London has been productive of much good. That laudable institution has been imitated here, and the *Bureau des Longitudes* was established by a decree of 7 messidor year III (25 June 1795). But this *Bureau* is on a more extensive scale, and endowed with greater authority than the Board of Longitude in England. The *Bureau des Longitudes* has under its inspection the National Observatory at Paris, and the one at the Military School, together with all the astronomical instruments belonging to the nation. It gives orders for the necessary regulations which take place at both the observatories; and appoints and pays astronomers and attendants. It employs itself in improving the astronomical tables, and the methods of determining the longitude, both by sea and land; in publishing the astronomical and meteorological observations; in calculating the *Connaissance des Temps*, which is published two years in advance, in order that the French navigators, when, on long voyages, they would determine the longitude, from the calculated distance of the moon from the sun or the stars, may be under no embarrassment. This, however, cannot happen at the present period; as the French have neither trade nor navigation. Privateers never proceed so far as to make the calculation of longitude necessary; for they take the first vessel they meet, whether she belong to friends or foes. The members of the Commission are men of the greatest celebrity: Geometricians, Lagrange and Laplace; Astronomers, Lalande, Messier, Méchain, and Delambre; Navigators, Borda and Fleurieu; Geographer, Buache; Artist, Carochez; Adjunct Astronomers, François Lalande and Bouvard. These commissioners meet regularly, once in every decade, and oftener when circumstances require it, in one of the smaller apartments of the National Institute. . . .

The Board of Geography (*Bureau du cadastre*) is a very good institution, and under the superintendance of the excellent Prony. The geographers employed by this Board are all taken from the Geographical School, and are therefore well acquainted with theory, and highly capable of performing all the mensurations and calculations relating to their department. Under the superintendance of this Board, are executed geographical and topographical admeasurements and descriptions of the territories of the Republic; geographical maps; maps for particular purposes, such as mines, forests, farms, inland navigation, &c; statistical calculations of the square contents and population of departments; population of cities, &c.

To this Board also belongs the calculation of new mathematical tables, according to the centesimal system (namely, one hundred degrees to a quadrant, one hundred minutes to a degree, one hundred seconds to a minute, &c.), both for the natural and the logarithmic sines, tangents, &c. According to some formulae published by Prony, the sine of any angle after 30° 10′ on the top of the page, is calculated by the differences 1, 2, 3, 4, &c. for every ten seconds, while the differences are invariably the same. The number of decimals, if my memory do not fail me, amounts to sixteen places. At the bottom of the page, the sine is in like manner calculated for the angle there situated, 30° 30′, with all its differences. By the regular method of adding the difference, one finds the sine of 30° 10′ 10″, of 30° 10′ 20″, of 30° 10′ 30″, &c. and can proceed all the way through, to 30° 30′, at the bottom of the page. By the differences, and their additions, the same result will be given to the lowermost angle at the bottom of the page, as has been calculated by Prony's formula of infinite series. By this method, those difficult calculations are made so perfect and simple, that any one can fill up occasional deficiences. Those tables are calculated by two persons, who compare their results; so that no error can remain undetected. The page is to contain the logarithms of sines, tangents, cosines, and cotangents, most of which are already calculated. Prony shewed me some stereotype proofs of the impressions of those tables, by Didot, inventor of that mode of printing. It will be very long before those laborious, expensive, and copious tables can be printed; but they will be the most complete and accurate that have ever appeared. Prony has consulted all his preceding labourors, Rheticus, Vlacq, Petifcus, Gardiner, Schultz, Vega, &c. In the new stereotypic edition of Callet's *Tables*, the sines are inserted after the new division of the quadrant into 100 degrees, but not into centesimal minutes.

Borda intends publishing new tables of sines, in centesimal degrees, minutes, and seconds, constructed in a commodious and perfect form, for finding out the seconds; nearly on the same plan as that on which Pezena and Callet's tables are constructed for 90 degrees, with sexagesimal divisions. On my visits to Borda, I have often found him occupied in correcting his tables. He complained that he could procure no paper, and must still defer the printing of them. Since the Revolution, manufactures of almost every kind, have either been stopped, or carried on very slowly. Money, that great spring which keeps the world in motion, has been wanting in their manufactures, commerce and every establishment whatever. In some parts of the country, the want of men has also been felt; for, when it is considered, that the government have, this last year, imposed a conscription of 200,000 men, besides from 4 to 600,000 already carrying arms, hands must necessarily be wanting in factories, which, even before this conscription, were at a stand.

It has been already shewn, that geographical maps, are under the superintendance of the Board of Geography. Cassini and De la Hire had, in the close of the last century, proceeded in measuring a part of the meridian of Paris; Cassini, the son and grandson, have since completed the rest of the meridian through France, and to this meridian have drawn a perpendicular. Finally, Cassini and La Caille, in 1740, repeated the whole measurement of the meridian through France. See *La Méridienne vérifiée par Cassini de Thury* (Paris, 1744). Since that time, the geographical situations of many sea-ports, towns and churches have been determined, and their latitudes and longitudes accurately ascertained. Besides drawing this meridian over a part of France, as a basis for an exact and general map, the Cassini's had it in view to complete these great triangles, by a general admeasurement of the situations of particular places, country towns, castles, enclosures, woods, roads, seas, rivers, sea-coasts, &c. and to publish an exact general map of France: But the necessary advances and subscriptions were wanting for purchasing instruments, and paying for the surveys, and for drawing, engraving, and publishing the maps. These extraordinary expenses, however, were defrayed, partly by the support of government, and partly by private contributions and loans. . . .*

The work was begun about the year 1740, and is at this time continued. These maps which, by the best informed geographers, are called the Cassinian maps, are one hundred and eighty-three in number, and form an atlas of France, so accurate and beautiful, that no other state whatever can produce a similar work. In the small kingdom of Denmark, the Academy of Sciences have at least imitated, if not surpassed, this excellent design: and it is with pleasure I reflect, that those geographical admeasurements were the principal labours of my youth, and are still carried on under my direction. The present Cassini and his associates had almost finished a general map of France, when the Revolution took place.

I have before observed, that Cassini was suspected of royalism and aristocracy. The ruling party seized the draughts of the admeasurements, drawings, and copper-plates, and even the innocent white paper belonging to Cassini, and deposited them altogether at the Board of Geography, where they still lie, and of which, for the present at least, no impression can be obtained; so that the Cassinian maps will, in future times, be a rare and scarce collection. Cassini complains bitterly on the subject, and has shewn me copies of several petitions to the government for reparation. It is possible that government may have good and sufficient reasons for preventing the circulation of these charts, while internal commotions are apprehended; but, on the other hand, equity and justice require that the property of Cassini and his associates should not be injured, and that the loss they have sustained should be made good.

*An account of these subscriptions and contracts is to be seen in: Cassini de Thury, *Description géométrique de la France,* Paris, 1783, pp. 194–200; *Projet et Acte d'Association pour l'entreprise d'une carte générale de France,* pp. 200–207; *Projet de souscription pour la carte de France en 173 feuilles.*

Chapter 4

TECHNOLOGY

It is not possible to justify any complete separation of pure and applied science in this period. Bugge himself did not attempt it. Yet it is convenient to collect together in one chapter several passages of the Danish astronomer's observations which concern industry and technology.

One of the most important events that took place during Bugge's visit to Paris was the staging of the first national exhibition of French industry. The Directory had decided that the standard of the products of French industry would be raised if specimens could be brought together into an exhibition, where comparisons could be drawn and awards given for merit. This would encourage competition. England had already shown how important industry could be to the prosperity of a country, and the Directory tried to convey a sense of occasion by fixing the date of the exhibition around the anniversary of the foundation of the Republic.

If this first exhibition was only a limited success, it was largely because insufficient notice had been given to the provinces. Instead, therefore, of being representative of the whole of France, nearly all the exhibitors came from Paris and the surrounding region (of the twelve prize-winners, nine came from the Seine department). The lessons learned from the first exhibition led to the unqualified success of the second exhibition, held in the last days of the Republican year IX (September 1801). Manufacturers were now fully conscious of the beneficial effects which publicity could have on their products, and this time the exhibition was not only on a much larger scale but was much more truly representative of the French nation. We must, therefore, in judging the first exhibition described here, appreciate its value in providing a precedent for the more justly famous and great industrial exhibitions of the nineteenth century.

Although in the sciences (particularly mathematics and chemistry), France was preeminent, this was not necessarily the case with technology. We should remember that, although the great political revolution had taken place in France, the Industrial Revolution occurred in Britain and it was helped by that new source of power, the steam engine. In the construction and development of steam engines, the French like the rest of the world looked continually for guidance to Britain. Bugge confirms that even the best French engineers were largely dependent on what they or their friends could pick up from visits to England. But the great engineer James Watt protected his inventions not only by patents but by preventing visitors from examining them at close quarters.

One of the perennial problems of Paris was the supply of water to all parts of the city. In the eighteenth century the pumps driven by water wheels in the Seine were no longer sufficient to provide all the water needed, so what Bugge calls "that important invention of the English," the steam engine, was introduced. James Watt had improved (1765–1768) on the Newcomen engine by devising a separate condenser, thus saving heat by allowing the working cylinder to be kept hot. Although this was an enormous step forward, Watt's single-acting engine still shared with the Newcomen engine the disadvantage of providing power only during the downward stroke and a heavy flywheel was necessary to carry the motion on smoothly through the second half of the cycle, when no power was produced. In 1782 Watt invented the double-acting engine, which involved a clever device for producing power during the second half of the cycle.

The most common use of steam engines in Britain in the eighteenth century was in pumping water from mines. By the end of the century rotative engines were used increasingly in the expanding cotton industry. Another important application was in the milling of corn: Boulton and Watt supplied engines to the Albion Flour Mills in London (1786, 1789, destroyed by fire in 1791); Jacques Constantin Perier (1742–1818) and his brother and collaborator Augustin Charles were not slow to follow their example in France.

An important task given to other steam engines in Paris at this time was the boring of cannon. This and Bugge's subsequent description of the gunpowder factory at Essones and the Arsenal serve as a reminder that France was at war. One of the remarkable features of Bugge's account is the full description of the manufacture of armaments and its publication in the midst of the revolutionary wars. Under the *ancien régime*, the extraction of saltpeter for gunpowder had been improved by the Régie des poudres. The great contribution of the Revolutionary period was vastly to improve the rate of production of gunpowder in response to the great demand. Technical improvements were also made in the manufacture of cannon.

The manufacture of two types of cannon is described in Bugge's account. Cannon made of cast iron, in order to be strong enough, had to be very thick. They were generally too heavy to be drawn by horses and were consequently used primarily by the navy rather than the army. The army preferred bronze cannon, as produced at the Arsenal.

Whereas there was, of course, nothing new in the eighteenth century about the use of cannon, the discovery of hydrogen ("inflammable air") and its application to flight were recent achievements. The French had been pioneers in the history of ballooning. On 4 June 1783, the Montgolfier brothers caused a hot air balloon to ascend nearly 3,000 feet at Annonay. Shortly afterward a balloon filled with hydrogen was successfully launched in Paris. The first human ascent was made in October of the same year by François Pilâtre de Rozier. Benjamin Franklin among others prophesied about the use of balloons in warfare but the realization had to wait for the French Revolutionary wars. Guyton de Morveau then suggested the military use of hydrogen balloons and on 7 September 1793, the Committee of Public Safety authorized expenditure of 3,000 livres for research on this. On 2 April 1794, after a balloon had been successfully launched at Meudon, the Committee of Public Safety gave orders for the immediate formation of a balloon company of twenty-five men under the command of a captain, eventually replaced by Conté. When this balloon company was sent to join the army, Guyton was instructed to accompany them and he had the satisfaction of witnessing a captive balloon successfully used for observation of the enemy's position before the battle of Fleurus (26 June 1794). The French victory at Fleurus was interpreted as a vindication of military ballooning and the next development was the setting up of a school for balloonists at Meudon, which Bugge describes here. Eventually, several difficulties, including the problem of mobility and the impossibility of advancing close to the position of the enemy, led to the abandonment of military balloons. The corps of balloonists was dissolved in March 1799.

There has been some debate as to how the hydrogen for the balloons was obtained. A modern authority states that it was produced by the action of sulfuric acid on iron.[1] Bugge's testimony makes it clear that, on the contrary, the hydrogen was produced by Lavoisier's method for the analysis of water, that is, the action of steam on red hot iron, and this is confirmed by another visitor, F. J. L. Meyer.[2]

Another invention of the Revolutionary period, which had potential military applications and of which the French were particularly proud, was the "telegraph." Although

[1]A. and N. L. Clow, *The Chemical Revolution*, London, 1952, p. 162.
[2]*Fragments sur Paris*, II, 115–117.

Bugge's principal comments on this are contained in his description of the Institute (see p. 84), it seems appropriate to mention the subject here. A visual telegraph system, in which semaphore signals were viewed by telescope from the highest point, had been suggested by Claude Chappe. The idea was submitted to the Legislative Assembly on 22 March 1792, but it was not until 5 April 1793 that the Convention appointed two of its members to examine the invention. Having realized its potentialities, the Committee of Public Safety decided that a line of signaling points should be established between Paris and the northeastern frontier, where the danger of invasion was greatest. On 15 August 1794, when the town of Quesnoy was recaptured by French troops, it took only one hour for the news to reach Paris from Lille, a distance of 150 miles. The Convention was suitably impressed by such rapidity of communications. Barère, presenting this achievement to the Convention, compared it with printing, gunpowder, and the compass, and asserted that it was an example of science serving "liberty."

Finally, no chapter on applied science in France in the post-Revolutionary period would be complete without a description of the Conservatoire des arts et métiers. Unfortunately Bugge's own account is disappointing. The account given by Henry Redhead Yorke, which is informative without being lengthy, is used here as a substitute. Although the English is contemporary with that of Bugge's translator, the change of author reveals itself in the invective against the commissioners who had appropriated works of artistic and scientific value in conquered territories.

The Exhibition of French Industry

The last two of the complementary days of every year are to be devoted to an exhibition of the different specimens of French manufactures, arts, and handicrafts, which are exposed to public inspection in a large building raised on porticoes or arches, in the Champ de Mars, opposite to the directorial amphitheatre. On the evening of the third complementary day, the Minister of the Interior, with the officers of the central bureau, reported the names of a jury appointed to examine, select, and pronounce on the best specimens in manufactures, arts, &c. which are deposited in those arcades for that purpose. As I have enjoyed a great deal of pleasure in loitering through those arcades, I shall present a short account of their contents:

1st Arcade. A pendulum which strikes decimal seconds, and shows the new divisions of time; the days are divided into ten hours, the hours into a hundred minutes, and the minutes into a hundred seconds.[3] This was executed by Couturier. On my return home, I chanced to alight on a watchmaker, in Colding, who had made a watch according to this new division of time....

2nd Arcade. Breguet, the famous watchmaker, has discovered a new escapement which is propelled by a constant and uniform force. This is a very lucky invention, and combines many advantages. Bruns, a carpenter, furnished many pieces of beautiful inlaid work in the cabinet line.

3rd Arcade. Fine razors, forged of steel, made in Clouet's new manner.

4th Arcade. Black lead pencils of different kinds, for designing and drawing lines, by Conté. They were of a peculiar composition, and superior to those of England.

5th Arcade. Different kinds of files, coarse and fine: they appeared to be very well finished.

6th Arcade. Desarnod's healthy and economic stoves, which are formed so as to consume a small quantity of fuel, and yet warm the room sufficiently....

7th Arcade. Different locks and scales. I could not discover any peculiar excellence in them.

8th Arcade. Some of the chemical and mineralogical productions of De la Place.

[3]The replacement of the traditional sexagesimal division of time by a decimal system is put in its context and explained by Bugge on p. 203.

9th Arcade. Several planetariums by Ruelle and Fortin; indifferent.

10th Arcade. Specimens of woven and printed tapestry, by Roby and Petit; both very fine, as well with respect to designs as colours.

11th Arcade. White earthenware manufactured by Potter, in the department of l'Oise; extremely fine and good.

12th Arcade. The model of a monument by Fouquet. . . .

13th Arcade. Fine specimens of cotton, carded and spun by the machines in Delaitre's manufactory in the department of Seine and Oise.

14th Arcade. Fine woven cotton, the produce of the manufactory of Fonfrède in the department of the Haute Garonne.

15th Arcade. Plain and printed cottons, the manufactory of Grémont and Barré; very fine.

16th Arcade. Different specimens of woven cotton, worth viewing.

17th and 18th Arcade. Excellent cloths of different colours, manufactured in Fère and Châteauroux.

19th Arcade. Pocket-handkerchiefs, the first specimen of the kind from a large manufactory, erected for that article in the department of Maine and Loire.

20th Arcade. All kind of smith-work, hatchets, spades, pickaxes, files, &c. Hardware, such as knives, scissors, snuffers, watch-chains; the polish fine.

21st Arcade. Specimens of woven cotton, the promise of improvement in that line.

22nd, 23rd, and 24th Arcades. Fine cloth called *Draps de Louviers,* manufactured in the department of Eure; not easy to determine which of the three should bear away the prize.

25th Arcade. Silk and cotton stockings, manufactured in Besançon.

26th Arcade. Cottons from Pont Audemer. The colours and patterns not very fine.

27th Arcade. Very excellent linen cloths from the same place.

28th Arcade. The finest pistols, rifle-barrelled guns, sabres of the most costly workmanship, the pride of the national manufactory at Versailles. These fine specimens of taste, invention, and execution, derived additional lustre from the manner in which they were grouped or arranged.

29th Arcade. Very fine patterns of tiffany and gauze.

30th and 31st Arcade. Fine specimens of tanned leather, from two tanneries in Pont Audemer.

32nd Arcade. Linen and pocket handkerchiefs from the same place.

33rd Arcade. Cotton stockings, and muslins from a manufactory in Troyes.

34th Arcade. Coppersmith work, excellent but still inferior to that of England.

35th, 36th, and 37th Arcades. Silk and cotton stockings, manufactured at Troyes.

38th Arcade. Sealing wax of different colours, scented. The sticks were very fine, and diffused an agreeable smell without being burned. I brought twelve sticks for thirty sous, each six inches in length, and a quarter of an inch in thickness, of different colours, red, white, green, and brown.

39th Arcade. Glasses of different kinds, blown at Gare, near Paris.

40th Arcade. A complete set of the new weights and measures, executed by order of the Minister of the Interior.

41st Arcade. The new weights and measures, executed by Ciceri, and in the 42nd arcade, the machines by which the new weights and measures are divided, by Kutsch; this artisan excels Ciceri in the execution of those articles.

43rd Arcade. Books printed on vellum paper in the office of Didot, the younger; namely *Contrat Social*, Juvenal, le Telemaque, Anacharsis, &c. They are all master-pieces in the typographic line.

44th, 45th, 46th, and 47th Arcades. Models of different machines. I could not find any marks of excellence in them, they were very clumsily executed. I was surprised to find that they should be offered as specimens of national ingenuity.

48th and 49th Arcade. Excellent specimens in general of cotton and woollen cloths, which did great credit to the factory in Beauvais.

50th Arcade. A large assortment of sabres, &c. manufactured by Provoteau.

51st Arcade. Plates of horn for lanthorns; very large, pure, and transparent.

52nd Arcade. Several costly articles of dress sewed in such a manner, that the seam was not to be discerned. I had not the good fortune to see them, as they were soon taken away.

53rd Arcade. Stoneware manufactured in Vaudrevanges in imitation of the English.

54th Arcade. Excellent tin work, such as ink-stands, flower-pots, &c. The form was beautiful, painted in different colours: some of the designs were very happily conceived and executed, they were done by Deharme.

55th Arcade. A handmill, well constructed, by Durand, which ground and sifted at the same time. . . .

56th and 57th Arcades. Several specimens of porcelain, the produce of the national factory at Sèvres, such as tea urns, basins, coffee-pots, plates, tureens, and large and small vases of all colours, figures, and groups, in biscuit, so white and fine, that they might be easily taken for gypsum. A round table of three feet in diameter, composed of many small pieces of blue ground, with white bas-relief, in imitation of Wedgwood ware; notwithstanding it was not free from blemishes, yet on the whole it was very neat and fine. On the porcelain there were two beautiful landscapes, fourteen inches in length, and ten in height. The form, designs, colours, and gilding of the porcelain at Sèvres are entitled to great praise.

In the mean time it may be proper to observe, that two kinds of porcelain are manufactured at Sèvres, soft and hard, the first is more showy, but the last approaches nearer to true and real porcelain.

58th Arcade. Pierre Didot, the printer, and Firmin Didot, and Louis Herhan, letter founders, or letter cutters, exhibited some of the newly invented stereotypic plates, in which each page of the book was cut or engraved, such as was used in the infancy of printing, but of a composition so hard, that it will serve to work off from eight to ten thousand copies. The expense is repaid in the number of copies, though they are sold at a low price. In this arcade I saw an edition of Virgil in 12mo. which sells for fifteen sous, Phaedrus for twelve sous, and *Fables* of La Fontaine for fifteen sous. There was likewise a splendid edition of Virgil on vellum paper, with copper plates, printed in this manner. A stereotypic edition of Callet's tables of logarithms, &c. Some books have just issued from the same press, which do great honour to this new invention.

59th Arcade. All kinds of crystal glass from Lebon's fabric in Creusot, in the department of Saône and Loire. These glasses are very beautiful in matter, form, and polish.

60th Arcade. A complete service of porcelain and decorations, designed for the table of a sugar-baker in Paris.

61st Arcade. The model of a threshing machine, by a miller in Rouen, not equal to our threshing machines in Denmark.

62nd, 63rd, and 64th Arcades. Spinning machines from a fabric in Luat, in the department of Seine and Oise, together with some sweet-meats by a confectioner in Paris.

65th Arcade. Porcelain from Dihl's and Gerhard's factory, Rue du Temple, Paris. This porcelain is better, and more durable than that of Sèvres; it is called in general *Porcelaine d'Augoulême*. Amongst many other fine pieces, I saw upwards of twenty paintings on porcelain, the largest of which was twelve inches long, and ten broad. The subjects flower and fruit pieces, a scene by moonlight, a young woman sitting, two old heads, and different landscapes. The designs were correct and natural, the colouring fine, the light and shade happily blended, and the execution of the whole inimitable. All these fine pieces did not experience any cast or blemish in the burning, which is not the case in other fabrics. It must be observed, however, that Dihl and Gerhard excel in the colour line, and that their furnaces are constructed in such a manner, that the colours do not melt or run into each other. The shades of the colours were much more delicate and clear than in those of Sèvres . . . .

66th Arcade. Many landscapes, designs in architecture, vases, and other figures engraved on copper plates by Defrance, which he calls "*tableaux en creux, gravés autour.*" This curious and excellent artist engraves the whole by a mechanical lathe, which imparts all the innumerable motions of the hand, and in many instances with greater success and perfection. With this instrument he can engrave plates after any design. He has also a factory of snuff-boxes of tortoise-shell, and other composition, for which he has found a very great demand. I expressed a wish to see his lathe, which I suppose to be a master-piece in mechanism; he assured me that no one ever saw it, except his wife or children.

67th Arcade. Perrin's metallic linen and gauze, or linen interwoven with steel threads. The texture beautiful, and of different degrees of fineness.

I was very much pleased with this new effort of the shuttle, and am persuaded this metallic tissue will be found extremely useful in many branches of manufacture, such as fitting in the porcelain, glass, and fine earthenware. It may also be employed in the making of vellum paper, and paper of different kinds, so as to render the transverse lines scarce visible. . . . The only articles in this collection that merited, in my opinion, peculiar distinction were the following:

Breguet's watch-work, particularly his escapement in the second arcade; Desarnod's economical stoves and boilers, in the sixth arcade; Berthier's steel-work in the twentieth arcade; firearms from Versailles, in the twenty-eight arcade; Grémont's and Barré's printed cottons, in the fifteenth arcade; Potter's white earthern or stone-ware, in the eleventh arcade; specimens of the stereotype and stereotypic printing, by Pierre Didot, Firmin Didot, and Louis Herhan, in the fifty-eight arcade. Defrance's mechanical engravings in the sixty-sixth arcade, and finally, Dihl's and Gerhard's porcelain, in the sixty-fifth arcade.

The whole exhibition, however, is very well worth viewing; besides, the idea is new. You there see many proofs of the industry and ingenuity of the nation. Whoever views them with an impartial eye must, however, acknowledge that they fall far short of that perfection of which they are capable; but when coupled with circumstances, such as when so many artists, manufacturers, &c. are enrolled as conscripts, or sent to the armies, the general scarcity of money, the want of encouragement, in consequence of the total extinction of trade and local convulsion, it is a matter of surprise that anything worthy of public attention could be offered. Let France once enjoy the blessings of peace: let the husbandman steer the plough in quiet, and reap the fruits of his own industry, then manufacturers, handicrafts, commerce and the fine arts, will daily gather strength, shoot forth, and expand into luxuriancy. Peace, I know, is the general wish of the people, a wish that is founded on good sense, and patriotism and industry will then contribute more to their real happiness, than the splendid but illusive acquisition of states and provinces, and the folly of disseminating the seeds of republicanism in other countries.

Steam Engines

[Two steam engines are used to pump water from the Seine, one at Chaillot and one at Gros-Caillou.] The Perier brothers, who are skilled engineers, built both of them. The engine at Chaillot is a single-acting Watt steam engine. The elder Perier saw the English engine in 1780 and afterwards built this one with expert skill. Prony has described and sketched it in his *Nouvelle architecture hydraulique* (*seconde partie*, Paris, 1796, pp. 86–99). Steam engines often need repairs and, so that the surrounding district should not suffer from lack of water during these times, two engines have been built side by side, only one of which is kept constantly in action. The water-tank and fire-grate in one of these steam engines were so dilapidated, that it could not work. While I was staying in Paris, the repairs to it had not even been begun. The water is brought by iron-pipes of about a foot in diameter for a distance of 150 *toises* to four brick reservoirs up on a hill; each reservoir contains 50,000 *muids* or 400,000 cubic feet.

The pipes are so arranged that the water can be brought into the particular reservoir in which the water-level has fallen owing to evaporation or has been lowered through use. From the reservoirs pipes lead to the various districts and sectors of the town. Once or twice each summer all the water is let out of the reservoirs and they are cleared and refilled; it takes ten hours to fill each of these. The whole thing is surrounded by a wall and a man lives on the hill whose business it is to see that the reservoirs are always full.

On the other side of the Seine between Rue St. Jean and Rue de la Boucherie stands the second steam engine at Gros-Caillou, likewise built by the Perier brothers and this too is a single-acting engine. It is distinguished from the engine at Chaillot only by being somewhat smaller and by some changes in the mechanism, whereby the steam is conducted into the lower part of the cylinder and pumped out from there. This engine is to be found, described and sketched, in the above-mentioned work of Prony (pp. 99–105).

The Perier brothers have also built a double-acting steam engine at a location in Paris called the *Île de Cygnes*. The purpose of this is the same as Albion's Mill in London, which was burnt down, namely to set millstones in motion and to grind corn. A very talented Spaniard, Bétancourt, travelled in France and England for several years at the expense of the King of Spain, in order to make a study of machines. He had so much money that he could pay not only draughtsmen to draw copies of the machines, but also craftsmen to construct good models. On his travels he collected many fine things and sent back to Spain many drawings and models which could be of use to his native land. He has now been ordered to return home and left Paris towards the end of August 1798; I, therefore, just had the opportunity of making his acquaintance and speaking to him a few times. Mr. Bétancourt went to London in 1788 to view Watt & Boulton's newly-invented engine. He also saw it working several times, but the inner mechanism, upon which the action really depended, was carefully kept from him and he was not allowed to examine it more closely. Therefore he went back to Paris and upon his return devised a way of constructing such an engine. He then had a working-model made on the scale of an inch to a foot or in the ratio of one to twelve. This model made a very good impression and was admired by the learned and by the mechanics and craftsmen of Paris. It was after this model that the Perier brothers built the first double-acting steam engine in Paris and set it up in a corn-mill. Prony has sketched and described this machine admirably (pp. 35–59 of his book).

The elder Perier is a member of the National Institute. It was there that I made his acquaintance. The two brothers went into partnership and built an excellent gun-foundry, in which iron cannon and other objects are cast; they also built fine, large workshops, where they make all kinds of machines. This large and important plant is not far from the big steam engine between Port aux Pierres on the Seine and the Grande Rue de Chaillot. In the foundry or rather outside it two reverberatory furnaces, five feet high on the inside, have been built. The furnaces discharge the molten iron into the foundry itself. It does not flow directly from the furnace into the casting-pit and the casting mould, but flows first into a very big, although shallow, iron crucible. Two such crucibles belong to each furnace. By means of a crane these crucibles filled with molten iron are brought to the aperture and the metal is then tipped into the casting-mould. In the casting-pit there lay four moulds, although more than one cannon is seldom cast at a time. The mould is made swiftly and well. It is of cast-iron four to six inches thick and can be taken apart lengthways. In each of these half-moulds sand is strewn and the model of a cannon in wood, turned according to the correct design and the exact size, is pressed into the sand. Both halves are then joined together by means of very strong bands, hoops, and bolts and then laid in the casting-pit. The whole thing is thoroughly dried out by fire and heat and all dampness dispelled. After this the mould is dammed up with sand. In this way Perier casts the largest pieces as easily as the smaller ones. I have seen iron cannons of larger and smaller calibre, cylinders, one to three feet in diameter, and pistons for steam engines, iron crucibles and pans four feet in diameter, conical iron wheels, driving-wheels one to five feet in diameter and many other objects of this sort, which were cast very well and without the slightest fault. If any molten iron remains in the crucible, bombs, howitzers, cannon-balls, and other smaller pieces are cast from it. Everything is cast so well in Perier's foundry that it would be hard to find better produced in the best English foundries.

The drill is driven by a small double-acting steam engine. The cylinder is one foot in diameter and four feet high. An epicyclic wheel is set in motion by the steam engine. This is really a kind of hinged rod, which is to be found in all steam engines, producing circular or rotatory motion. In order to maintain the smoothness of the rotation a large iron flywheel about ten feet in diameter is joined to the epicyclic wheel. In this way the main wheel is set in motion, turning four cannons by means of several wheels connected to it. The knob of the cannon is fastened to the axle of the wheel; the cannon lies supported in two places on shaped pieces of wood, one near the gun-support and the other near the muzzle. The drill lies horizontally and is fastened to an iron-rod, the cogs of which engage a driving wheel and on the axle of the latter is a larger wheel. This can be set in motion, either by means of weights or by hand and by this the drill is made to bore deeper and deeper into the heart of the cannon. First the cannon is hollowed out by smaller and afterwards larger drills and finally the barrel is completely smoothed and polished by the last drill, which corresponds exactly to the required calibre of the cannon. When the cannon is drilled the hoops at the muzzle, middle and end are also turned. This method results in the complete rounding and correct geometric shape of the cannon. It is clear that all the hoops and bands must be concentric with the axis of the bore. The positioning and aiming of such a cannon will therefore be perfectly accurate and reliable and its shots sure.

At the other end of the plant is another steam engine, which similarly drills four cannons, so that eight cannons can be perfectly drilled and shaped at the same time.

According to an agreement which they have entered into, the Perier brothers produce iron cannons, howitzers, mortars, bombs, and bullets for the Republic.

I saw here a good way of destroying old iron cannons and casting new ones from them. Near the drill and close to the foundry a strong beam was erected at an angle of about thirty degrees. At the top end was a strong pulley which raised a heavy iron weight. Vertically below this pulley the cannon was laid on the sharp protruding edges of two strong pieces of iron. Then the iron weight was dropped and it struck the cannon directly between the two supporting pieces and thereby smashed it into two, three, or more pieces.

Besides this fine gun-foundry the Perier brothers have built a machine-factory at the same place. It consists of a main-building and two large wings. In the large building is the warehouse for raw materials and finished machines. In the wing jutting out towards the Seine on the ground floor are the smithy, turner's workshop, and the metal-work shop; on the first floor is the carpenter's shop where the machines are made and fitted together.

I saw various machines for combing and spinning cotton—some finished, some under construction. The iron-chain which goes from the main wheel to the smaller ones is made by a machine, so too are the cogs of all the metal wheels. The old well-known methods are used here as well as new ones. In the other wing is a small steam engine which can be used for forging heavy pieces of iron, cutting screws, and shaping large pieces of iron or metal.

These splendid engineering premises are the private property of the Perier brothers. In their house in another part of Paris they have a large office where they conduct their extensive business and deal with the accounts connected with it.

In cupboards and drawers in the office is a considerable number of designs for machines. They told me that they produce for private individuals for a fair and agreed sum of money, steam engines, combing-, spinning-machines and all other known machines, even completely new ones, provided that accurate designs, plans and sketches are submitted to them.

The School of Military Ballooning

The Aerostatic School in Meudon was established by a decree of the Committee of Public Safety, of 31 October 1794. This school consists of director, sub-director, a secretary, a magazine-keeper, and sixty pupils, who are instructed in all that relates to the aerostatic science, especially such parts of it as may be directed to military operations. There are two rooms set apart in the old castle, for the construction of the air balloons, with all the apparatus necessary for that purpose. The pupils, with Conté the director, lodge in the new castle. M. Conté is an able natural philosopher:[4] he cannot be too highly praised for his unremitting attention to the regulations and management of the School. He is well known for his inventions, such as the aerostatic telegraph, and his factitious black lead pencils, which are brought to such a degree of perfection, as to rival the best in England: they are not prepared from the native ore, but a composition which consists, as far as I have learned, of iron and sulphur.[5]

The balloons in Meudon are made of a peculiar kind of thick taffeta, woven for that purpose. When sewed they are varnished over so that the pores are closed in such a manner as to prevent the evaporation of the gas in a very considerable degree, which is the reason that those balloons hold the hydrogen, or inflammable air, many months; whilst others that are not prepared in the same manner are found to be exhausted in a few days. To the improvement of the gas, M. Conté has not a little contributed to the manner of filling the air balloons. The mode is to erect a small furnace, through which several large iron pipes pass (commonly from four to six), which are filled with iron filings. The ends of these tubes extend out of the furnace, and are furnished with a cock, which may be opened or shut at pleasure. A smaller tube is joined to the end of these pipes, and is then inserted in the lid of the copper or vessel, half filled with water, and so air tight that the steam can only find its way through the red hot tubes in the furnace. From the opposite end of these red hot tubes, which

[4]"an able physician as well as a chemist"—J. J.

[5]N. J. Conté patented a process in 1795 which has since been adopted generally for the manufacture of lead pencils. Cut off by the war from the traditional Cumberland graphite, Conté experimented with poorer quality graphite, which he pulverized and mixed with purified clay into a paste. This was squeezed into a thread, which, on drying became the "lead" of the pencil. Bugge's suggestion of the composition of this "lead" was therefore quite mistaken.

run out of the furnace, a small tube goes into the vessel, which is filled with a solution of caustic lye, or alkali, and then it passes to the tube which conveys the hydrogen gas into the balloon.

The whole apparatus, cauldron, furnace, &c. may be erected and worked in two days. A balloon of about thirty feet diameter may be filled in two or three days. When a balloon of this size is newly filled, it will carry up a weight of 2,000 pounds, and twenty men at least. In two months it loses so much by evaporation, that it will only bear 500 pounds, and ten men. I have seen the experiment tried in the Champ de Mars on the feast of the New Year, in the seventh year of the Republic. Such balloons are always found ready filled on the terrace at Meudon, where they stand in the open air without receiving any apparent injury, in consequence of the peculiar texture of the taffeta, and the excellence of the varnish. The upper part was covered with a coat or case of fine leather, from whence the ropes descended, to which the car was attached. All these military balloons are tied together, and aerostatic soldiers are taught to manage them.

In mild or serene weather a number of these soldiers ascend, always accompanied by an officer or subaltern. Two companies of aerostatic soldiers are always quartered at Meudon. Each consists of one captain, two lieutenants, two serjeants, two corporals, one drummer, and forty privates.

The Powder Factory at Essones and the Arsenal

At the beginning of the Revolution when war was the predominant concern, the gunpowder plant at Essones was reorganised and it was thought that a far greater quantity of gunpowder could be prepared by a new method of manufacture. It is not manufactured by pounding the ingredients. A very large wheel with cogs turns twelve cylindrical containers by means of driving wheels attached to them. These receptacles have a very strong base, but are open at the top. Into these is shaken the previously ground quantities of saltpetre (which is only used here in refined form), sulphur, and charcoal; and on top are laid metal balls. By the movement of the machine and the continual revolution of these receptacles round their vertical axes the quantities are pulverised and mixed. When it is fine enough, it is taken out of the round receptacles and laid on a coarse and wet cloth about two inches thick and then on top of that again another wet cloth and a new layer of the mixture is placed and so on. After this the whole thing is enclosed in frames and boards, so that nothing is lost and then put under a press with a strong iron screw. By means of heavy pressure the mass is completely soaked through and made into a paste. When it is taken from the pressing cloths, it is put into sieve in which there is a strong wooden roller and as it rolls backwards and forwards the mass is pressed through the holes of the sieve and at the same time granulated. The movement of the sieve is not brought about by any mechanism, but by hand.

The ratio of the parts is as follows: for 100 pounds of gunpowder, 66 pounds of saltpetre, 25 pounds of charcoal, and 9 pounds of sulphur are used. However, after some new experiments had been made, it was decided to reduce the quantity of charcoal and increase the saltpetre. There is no doubt that this powder must be very powerful and effective, as only well-refined or purified saltpetre is used in its manufacture. By use of this therefore a large quantity of elastic gas must be disengaged and on this alone depends the initial speed and force of the bullet. According to the report of the well-known authority on artillery, St. Remy, in most of the gunpowder mills in France and in the best ones, for a hundred pounds of powder, 76 pounds of saltpetre, 12 pounds of sulphur, and 12 pounds of carbon used to be taken. Doubtless, now, on account of the purity of the saltpetre, it has been possible to reduce the quantity without spoiling the strength of the gunpowder and, moreover, there is the advantage of being able to sell it more cheaply now than before. Previously in the ratios and strength of the parts a distinction had been made between gunpowder for mortars, cannons, and muskets. Experience, however, has taught that such a distinction causes more harm than good and now all the gunpowder that is used in the field is manufactured to the same strength and in the same ratio.

In the gunpowder-mill at Essones as I have already said, the ingredients are not mixed by pounding and made into a paste, but the saltpetre, sulphur, and carbon is reduced to a powder by rolling metal balls amongst it in containers and the paste is formed by pressing it between wet cloths. This method of manufacture is quicker than the usual pounding; however the particles do not appear to be so well mixed and thus the grains of powder can apparently become hard and firm. I am told also that Essones powder turns to dust more easily than the usual sort.

In the Arsenal, which is situated in the suburb, St. Antoine-sur-Seine, is the Republic's own gun-foundry, where metal cannons are expertly cast. For gunmetal ten parts copper and one part tin are always used. The copper must be completely liquid before the first has melted. After the mixing everything must be thoroughly stirred so that the lighter metal does not float to the top. Finally, before casting, the mixture is subjected for a short time to an intense heat. Monge has given a detailed and excellent description of everything concerning the preparation of the moulds and the casting. The drilling-machine is laid horizontal on a strong float and is driven by the current of the Seine. This drill has also been described in various books.

Neither in France nor in England is any mystery made of the construction of a drilling-machine. When in 1777 I met the English artillery-captain, Smith, I was given permission to visit the foundry and drilling-machine as often as I wished. The latter similarly is horizontal. The cannon is turned on its axis by means of a horse-mill; the master-driller presses the borer in. This has a driving wheel connected to cogs on its shaft. This drilling device was very good and accurate. By four hanging weights, two above the cannon and two above the drill, you could see immediately whether the borer and the axle of the cannon formed a straight line and if the cannon was being drilled quite straight. I have not seen such a contrivance in the French drilling-plants; but I have no doubt at all that they have other sure mechanical means of boring straight through the cannons. In Holland and in some places in Germany a great mystery is made of drilling-plants; even in my native country, I was not able to obtain permission to inspect the drilling-plant at Friedrichswerk. But it is basically only a strange caprice of the master-driller, wanting to make a secret of a matter which is so well known and has been so often described.

The Conservatoire des arts et métiers (described by Yorke[6])

Through the indefatigable exertions of Bishop Grégoire, the National Convention, on 1 October 1794, decreed the establishment of a Conservatory of Arts, whose object was to collect machines, utensils, designs, descriptions, and experiments relative to the improvement of industry, and to diffuse the knowledge of them throughout the Republic. But it was one thing to decree, and another to execute. By an inconceivable remissness, the execution of that law was suspended for three years. National edifices were often granted by dint of favour to useless projectors, while the Conservatory of Arts could not obtain a place wherein to display its riches and means of instruction. Grégoire has repeatedly told me, that a crowd of respectable artisans, supposing that as he was the reporter in the business, he could accelerate the establishment of the institution, have often applied to him with the bitterest complaints, that everything was done for the agreeable arts, but nothing for those which are useful.

On 16 December 1795, the Executive Directory sent a message to the Council of Five Hundred, desiring that a portion of the former Abbey of St. Martin should be appropriated to this object. Almost a year after this message, the Council resolved that no other expense should be incurred for the Conservatory than what was necessary to prevent the instruments from going to decay. This magnanimous, economical, politic, and philosophical decree, by which a few pounds were saved, by suffering what had cost thousands to go to ruin, was soon called in question; because, for want of a proper depot, the immense and invaluable quantity of objects which had been accumulated, could not be preserved; and the funds necessary to set the institution in motion, would

[6] *Letters from France*, II, 29–35.

prove to be laid out to the greatest interest from the influence which it would have over national industry.

At length a decree passed on 7 May 1797, by which the three depots where the instruments had been distributed; namely, at the Louvre, in the Rue Charonne, and in the Rue de l'Université, were united into one, at the *ci-devant* abbey, St. Martin des Champs, in which the municipal administration of the fifth juridical division of Paris transacts its business; and the sum of fifty-six thousand nine hundred livres was appropriated for the reparations of the building, the purchase of the land, and the indemnity accorded to the renter.

Thus finally organised, the Conservatory of Arts presents a splendid accumulation of useful machines, always open for the inspection and improvement of the public. According to the plan of the institution, it contains, or should contain, all the instruments of those arts, by the help of which men may nourish, clothe, lodge, and defend themselves; and it maintains a correspondence with all parts of the world. The machines which Pajot d'Onsenbray gave to the ancient Academy of Sciences, and those which were added to them by that learned body, as well as the greater part of the beautiful models which composed the gallery of mechanical arts, belonging to the late Duke of Orleans, are now all collected in the Conservatory. Besides these, there are above five hundred machines, bequeathed to the government in 1783, by the celebrated Vaucanson, to whom the French nation is as much indebted as to Olivier de Serres, and Bernard Palissy, the fathers of French agriculture and chemistry. The collection of Vaucanson comprises many ingenious machines for the preparation of threading materials, for carding and spinning cotton, twisting silk, and all kinds of weaving; shuttles for ribbands and lace; instruments for knitting, for stuffs of different colours, and for fabricating at the same time several pieces in the same loom. These models have already multiplied the number of cotton spinners. One of these machines, which Vaucanson invented out of pique against the Lyonese, is remarkable for its singularity. An ass, by turning a capstan, set in motion the shuttles and every part of the loom, and manufactured a drugget with flowers, a pattern of which has been preserved. Here are, also, the tools which Vaucanson used in the construction of his machines. The one employed for making iron chains is so simple, that a workman, in less than half an hour, may begin to use it. The strength of man is increased an hundred fold by such inventions.

In addition to these collections, there is an immense number of machines relative to agricultural labours, such as draining, irrigation, preparation of oil, according to the Dutch process, &c. &c.; also the ingenious machines with which paper money has been fabricated, among which, is the mechanical arithmetician or marker, of Richer, which, by a single motion of an entire printing-press, performs all the changes of numbers, in the natural order of the cyphers, from 1 to 9999.

The Conservatory contains also machines for twisting tobacco, which were taken from on board English vessels, as well as a very important chart of the coast of North America, executed by order of our [that is, the British] government. It has been also enriched by the discoveries of those French *savans*, those learned robbers of the National Institute, who composed the board of plunderers, that followed the victorious marches of the Republican armies in Holland and Italy. Whole waggon-loads of instruments of husbandry have been filched from their proprietors, and transmitted to this National Reservoir, by those industrious and indefatigable literary thieves, citizens Thouin, Faujas, Leblond, Berthollet, Barthélemy, Monge, Moitte, and De Wailly.

Among the more recent inventions, are, the whitening of linen by oxygenous muriatic acid,[7] by Berthollet; the manufacture of minimum, by Oliver; Seguin's method of preparing leather in a few days, which formerly underwent a preparation of two years; and a barometer, by Conté, one of the Conservators, wherein the weight alone of the mercury serves to denote, with great precision, that of the atmosphere.

The mechanical part of the arts and handicrafts, the construction of machines, and the best finished utensils, their mode of acting, the distribution and combination of their motion, and the

[7]Chlorine.

employment of their power, are explained by Le Roy, Molard, and Conté, who are the Conservators; and the designs are executed by Beuvelot. This method of instruction, with the models at hand, is preferable to a thousand didactic lectures.

It is intended to add to the machines, specimens of the produce of French and foreign manufactures, in order that they may be compared together; a drawing of each machine, and a description which may perpetuate the idea of the inventor. A vocabulary, and a reference to those works which have treated upon each subject, will be also published, in order that the various denominations, given to the same thing in different parts of the Republic, may be reduced into one system.

The object of the Conservatory is not only to secure to the public the knowledge of those inventions for which the government has conferred rewards or granted patents, but also to become the common depot of all the inventions in the arts. Thus, it is for the useful arts, what the Louvre is for sculpture and painting. The defect of communication between the different parts of France, estranged them, in some measure, from each other, and hindered the circulation of useful proceedings; another object, therefore, of this institution is, to transmit to the departments an account of every invention or improvement which may extend the branches of industry, by abridging the labours of man.

Upon the whole; the Conservatory of Arts is one of the most beneficial, and most laudable establishments in France. It has a direct tendency to stimulate genius, and encourage industry, and its effects upon Agriculture, Commerce, and National Prosperity are incalculable.

Chapter 5

SCIENCE—PUBLIC AND PRIVATE

One of the features of life under the First Republic that most impressed Bugge was the extent of the patronage of science by the French government. Yet although there were many institutions controlled, staffed, and financed by the central government, there was still room for private individuals such as the entrepreneur Desaudray or the physicist Charles to propagate science in their own way. The makers of scientific instruments described in this chapter were their own masters, but they now depended to an increasing extent on orders from state institutions. In this chapter, therefore, we begin with details of the munificence of the public purse and then pass to the efforts of private societies and individuals, which are thus seen in perspective.

Bugge's account contains a most interesting summary of the budget of the Ministry of the Interior for the Republican year VII (1798–1799). It is impossible to find a single modern equivalent to express the wide responsibilities of the Minister of the Interior. He was not responsible for police, but there were few other internal aspects of the modern state which did not come within his responsibility: public works, state factories, education, science, the arts, prisons, hospitals, support of the poor, and many more. In reproducing this budget, there is a temptation to try to calculate what fraction of the total forty-four million francs was devoted to the support of science as distinct from other activities. Section V is obviously the one most relevant to science and 4,951,103 francs was allocated for this section. This total, however, included expenditure on technology and the arts. On the other hand, not included in this total was the expenditure on the office of weights and measures, the school of highways and bridges, and the school of mines. In any detailed evaluation of government support of science, one would also have to look to the state factories such as that at Sèvres.

Bugge was obviously impressed by the large financial support given to scientific institutions by the French state. It formed an obvious contrast to the situation of other countries, even powerful and wealthy states such as Great Britain. The leading British scientific society, the Royal Society of London, was linked with the head of state in name only. The only new and important scientific institution in Britain in the years following the French Revolution was the Royal Institution, but again its name was misleading, being supported entirely by private subscription. But if it is permissible to make the generalization that in Britain science depended for its advancement on the interest and patronage of private individuals, it cannot be said that in France it depended entirely on the state. Although most of the great institutions with international reputations had state support, there were a number of societies and even some meritorious educational institutions which were in private hands. Although several quite worthless cultural societies sprang up in Paris under the Directory, there were some which had distinguished members and did serious work. Bugge was particularly impressed by the Natural History Society. A private society of more lasting importance was the Philomatic Society. Although its more eminent members seldom took an active part in its work, it provided a useful forum for the presentation and discussion of scientific papers. Because of the general exclusion of younger men from the official body of French science, the Institute, the Philomatic Society was particularly valuable to those who had not yet established their reputation.

In the field of higher education, private enterprise

could hardly compete with state patronage. In the sphere of popular education, however, the Lycée républicain made a valuable contribution. Founded originally in 1781 as the Musée, it became successively the Lycée, the Lycée républicain, and the Athénée de Paris. Like the Royal Institution in London, it was supported by private subscription and the lectures were aimed at a wide audience, including ladies. It was capable, as Bugge observed, of communicating a taste for science rather than any deep knowledge. The private subscriptions were not always sufficient to pay the salaries of the professors and the Lycée applied in 1792 and 1793 for financial assistance from the government. Another society supported by private enterprise but looking, like the Lycée républicain (with which it is often confused), to occasional government support, was the Lycée des arts. Originally founded in 1792 by Charles Gaullard Desaudray primarily as an educational institution, it developed into a scientific society. Although the Lycée des arts was concerned with the practical applications of science, its interests went beyond this, as the reports on the meetings attended by Bugge show.

The most outstanding private course in science was undoubtedly that given by the physicist Charles, who thus carried on a tradition of private teaching from the *ancien régime*. Apart from Charles's innate ability as a lecturer and a man of science, the secret of his success depended on the fact that his approach to the subject was completely experimental and that his extensive collection of physical instruments, built up over a lifetime and jealously preserved, was unrivaled even by official institutions like the École polytechnique which had benefited so extensively from the Revolutionary confiscations. Charles had a large following because he deliberately avoided taxing his audience with technical terms and mathematical calculations. He relied instead on what he could demonstrate and this visual appeal brought wide popularity.

Finally Bugge visited and described the activities of the Paris instrument makers, the skilled artisans on whose craft so much work in physics and astronomy was dependent. This is traditionally a neglected part of the scientific spectrum and the Danish astronomer rescues from obscurity men like Dumotiez and Laroche.

State expenditure of the Minister of the Interior for the Year VII

By the law passed on 11 brumaire year VII (1 November 1798), the Minister of the Interior is allotted the sum of 44,143,374 francs. As this provides evidence of how much the Republic is actually spending, or at least intends to spend, on public works, science, art, manufactures, agriculture, &c., I am including an extract from this law:[1]

Ordinary and Regular Expenditure

SECTION ONE	
Salaries of commissioners in the central and municipal administration	3,720,100 francs
Salaries of *concièrges, officiers de santé,* &c.	1,200,000
Food, clothing, bedding, linen of prisoners, the indigent, &c.	3,850,000
SECTION TWO	
Hospitals and charitable institutions	17,061,600
SECTION THREE	
Planning and maintenance of canals	3,500,000
National and civic buildings	1,500,000
School for bridges and highways	72,000

[1]Although this section is based on Bugge's account, slightly more detail is given here.

Council for mines. Inspection and School of Mines	245,610 francs

SECTION FOUR

Instruction in rural districts	45,000
Orangery, garden and plantation at Versailles	31,000
Plantation at Rouille	10,000
Stud at Rosières, depots at Le Pin, Bayeux, Pompadour, stallions, &c.	206,961f. 25c.
Veterinary schools	290,000
Encouragement of agriculture	400,000
Sèvres factory	100,000
Gobelins factory	180,000
La Savonnerie factory	40,000
Beauvais factory	48,000
Encouragement of arts and crafts	200,000
Premiums for general subsistence	100,000

SECTION FIVE

National Institute: salaries, 280,000f. Travel expenses for tours, 120,000f. (As far as I know these journeys have not yet been undertaken.)	400,000
École polytechnique	394,133
Medical school in Paris	266,972
Medical school at Montpellier	148,752
Medical school at Strasbourg	80,480
Board of Longitude	100,533
Observatory	10,000
Collège de France	99,829
Prytanée français (for orphans)	119,082
Liancourt's school	354,000
School of painting, sculpture, and architecture	80,188
School at Rome	34,950
David's school	2,400
Free school of drawing	20,600
Central museum of art and other expenses of the National Palace of Science and Art	112,410

Museum of French monuments	34,920 francs
French school of painting at Versailles	44,980
Schools of painting, sculpture, &c. in the provinces	38,000
National History Museum and botanical garden	269,578
Collection of minerals at the Mint	14,800
Vaucanson's collection of machines	11,920
Museum at Versailles	9,920
School of mechanics and *Conservatoire des arts et métiers*	119,800
National Library	149,413
Lectures on oriental languages	23,000
Library of the Four Nations	27,660
Library of the Arsenal	37,820
Library of the Pantheon	32,400
Museums in the provinces	41,075
Book collections	76,270
Buying books, paintings, busts, medals, manuscripts for various collections	100,000
Conservatory of music	309,496
Opera house (*Théatre des arts*)	250,000
Preparation of the national survey map; trigonometrical tables	119,000
Geodesic work; measurement of the arc of the meridian	99,000
Map of France	25,000
Telegraph	235,492
Aerostatic school at Meudon	31,230
Riding school at Caen	6,000
National festivals	400,000
Encouragement of writers	200,000

SECTION SIX

Expenses of the Minister and his staff and incidental and miscellaneous expenses	1,082,000

Exceptional Expenditure

SECTION ONE	
Building and improvements of prisons and court houses	1,500,000 francs
SECTION TWO	
Assistance for refugees from the colonies, &c.	1,000,000
SECTION THREE	
Reorganization of educational establishments	100,000
SECTION FOUR	
Farm at Sceaux, destruction of wolves, buying stallions	890,000
Subsidies for clock factories at Besançon, Versailles, and Grenoble	120,000
SECTION FIVE	
Office of weights and measures	120,000
Construction of weights and measures as standards and as models to be sent to the departments	1,000,000
Completion of the Museum of Natural History	150,000
Completion of the Central Museum of Art	200,000
Transport of the new monuments	200,000
Engraving of plates for the *Journey to Syria*	30,000
SECTION SIX	
Office to deal with outstanding business of the Minister	20,000
TOTAL of ordinary and exceptional expenditure	44,143,374f. 25c.

All these sums required by the Minister of the Interior for the expenses of the year VII amount to 44,143,000 francs or eleven million imperial thalers and therefore exceed by almost half the annual state income of Denmark.

The expenditure on hospitals and the care of the poor is

considerable (over seventeen million francs). The Minister of the Interior and the offices under him cost more than eight million francs. I consider the estimate of 120,000 francs a year for the Republican office of weights and measures and a million francs a year for the preparation of standards and models very high. More than 240,000 francs are allotted to surveying and cartography. The Board of Longitude has an income of 100,000 francs a year, whereas the Observatory has only the small sum of 10,000 francs.

The fifth section of ordinary expenditure shows how well provided for are science and the arts, and what large sums are being spent on everything relating to public instruction, particularly on practical subjects useful to the state. All this is excellent, significant, and generous and proves how highly the present French government regards science and how much it is endeavouring to blot out and expiate the disgraceful crimes of vandalism against the Muses.

From the above list it may further be seen that, with the exception of the two schools of medicine at Montpellier and Strasbourg, the large sums mentioned are to be applied only in Paris for the benefit and promotion of science and the arts and for the instruction and enlightenment of the public. The resultant benefits are therefore to be enjoyed only by the Parisians or the few clever persons who can find an excuse for coming to Paris—and how small their number must be, against 27 million.

In France there are 104 towns, each of which has over 10,000 inhabitants. Apart from 98 central schools (of which only two-thirds have been organised), it seems according to this list that not much has been done for the promotion of learning and education amongst the numerous inhabitants of so many large towns. Would it not perhaps be more to the purpose to have smaller and less magnificent scientific institutions in Paris and divide some of these riches amongst the large and important towns in the departments? A house is always better lighted, if the lights are distributed throughout the rooms, than if they are gathered together in one large hall.

Private societies

Besides the large number of public educational establishments, which I have mentioned in previous letters, there are even more private societies in Paris, the purpose of which is to entertain themselves with literary occupations and at the same time to spread enlightenment and disseminate knowledge in a comprehensible and popular fashion, even amongst the gentle sex, which diligently takes part in them.

These private societies are the following: *Société libre des sciences, lettres et arts*—in the national palace; *Société des amis des arts; Société de belles lettres; Société d'histoire naturelle*—in Rue Anjou-Thionville; *Société d'agriculture; Société polytechnique; Société philomatique*—in Rue Anjou-Thionville; *Société de médecine*—in the national palace; *Société médicale d'émulation*—in the School of Medicine; *Société libre d'institution; Lycée des arts*—in the *Palais d'égalité* or the former *Palais royal; Lycée républicain*—in the Rue St. Honoré; *Lycée des étrangers*—in the Rue Cerutti.

The Free Society of Science and Art holds four public meetings annually; in one of them in summer 1798 the following were read: some passages from Ossian, translated by Miges, a treatise on Petronius of Deguerle, two fables of Robert; "Is the French Revolution beneficial or harmful to literature?" by Thiébault, some poetical pieces by Boutet, Demachy, and Simon. By this can be seen that history and *belles lettres* are the purpose and occupation of this society.

The Society of Friends of Art (*Societé des amis des arts*) consists of artists, connoisseurs, and art-lovers, each of whom pays a yearly contribution. With the sum received paintings, etchings, and statues are bought. At the end of the year lots are drawn for these and it can easily happen that for a small contribution some members receive very good and valuable objects. Talented artists can sell something of their work to this society. The society has suggested to the government that it should order each department to take at least two of these lottery tickets in the hope of winning some masterpieces of French artists and adorning their towns with them. "Also," in the words of the printed proposal, "the capitals of the departments would then no longer complain, that everything is done for Paris and nothing for them; that everything that could arouse the curiosity of the whole world is being collected in Paris, while the departments are ignorant of art and science and altogether deprived of their products."

The Natural History Society has done many useful things particularly at the time of the Revolution. When science was regarded with contempt and all instruction came to a standstill, this society was the only one which continued its lectures in all fields connected with the very useful knowledge of nature and by its prizes continued to encourage diligence and talent. It has published a volume of papers read to the Society (*Actes de la société d'histoire naturelle*). In order to have a still greater and wider influence it was reorganised a year ago. The meetings take place now only once a month, but the society has elected an administrative body which meets once every ten days. Again this body has chosen six commissioners—Jussieu, Lamarck, Haüy, Fourcroy, Desfontaines, Lacépède—and assistants or adjuncts—Ventenat, Brongniart, Vauquelin, Cels, and Millin. Cuvier is general-secretary and Silvestre treasurer. The Natural History Society is, moreover, divided into two sections: the members of the first have undertaken to give a lecture once a year. The members of the second are bound by no definite work and are called *associés libres*. This winter (1798–1799) Cuvier delivered a lecture on comparative anatomy about the difference in brain structure of warm-blooded animals. It received a great deal of praise from the experts. The administration is busy editing the lectures which have been given and the first part has already been printed: I consider that the Natural History Society is one of the best-organised, most active, and most useful of the private societies in Paris. Its aim is not only entertainment, but also a zealous furtherance of science and as long as such famous men, eminent and active in their own fields, participate, this aim must certainly be attained. . . .

The Philomatic Society has a very extended plan, which embraces almost all the sciences. At the beginning of the year

1798 the secretary, Silvestre, published a general report of its work from 1 January 1792 until 1 January 1798. He declared that this society had achieved much in the time which had elapsed between the suppression of the former Academy and the establishment of the National Institute. Then he gave a list of the lectures which had been given in the last six years and regretted that the society had lost three of its most promising young members, Bonnard, Vié, and Riche. The latter died on one of the ships which had been sent to look for La Pérouse. Cuvier wrote a fine panegyric in his honour. This society publishes a journal (*Bulletin*) in which it acquaints its members, particularly those in the provinces, with new writings, discoveries, and experiments. It is part of the plan of this society to repeat all new and important experiments and to affirm or reject the ones which might be uncertain.

In the summer of 1798 and at the beginning of 1799 the following papers were given in this society: Tonnelier—observations on various volcanic products. He spoke of pearl-stone, the volcanic zeolit, which Assessor Esmark described in his mineralogical journey through Hungary; Latreville spoke about the ant; Dumeril about the osteology of birds; Chaptal about a cancerous disease in the wood of chestnut trees, which is cured by making charcoal of the surface of the damaged part; Vauquelin spoke of potter's clay and pottery.

The Medical Society (*Société de médecine*) was very famous even before the Revolution and had published several volumes of its memoirs. After the Revolution it was reorganised and since then has published five volumes of its writings (*Recueil la Société de médecine de Paris*). Every month one issue consisting of five or six sheets appears, numbered in such a way that a year's set or twelve parts form two volumes.

The rival medical society (*Société médicale d'émulation*) really just provides an opportunity for beginners to practise this science. Under the supervision of their teachers and other famous doctors and surgeons they deal with various subjects. Two volumes of these papers have been published (*Mémoires de la Société médicale d'émulation*).

The Free Society for Education (*Société libre d'institution*) seeks to improve the initial education of youth and to make it easier. At its public meeting on 6 brumaire year VII (27 October 1798) the following lectures were given: Harger on the best method of teaching writing; Arnaud and Le Blond made a report on Couret's hand-book of morality and politics (*Manuel de morale et politique*); Bouillat read out some lines of poetry about the floods caused by the River Adige; Roulle spoke of the necessity for better teaching methods; Le Blond spoke of ways of denoting decimals and tens; since many people now use a comma to distinguish thousands, millions, billions, &c., the same thing, in his opinion, should not be used to distinguish whole numbers from decimals. He on the other hand suggested the use of the question mark instead, so that 620? 345 would mean 620 units $\frac{345}{1000}$. But this absurd method of notation will presumably not be adopted by mathematicians. Couret read "The Throne of Snow," a fable; Mercier gave a report on Valain's letters on the art of writing; Le Blond spoke of learning by immediate perception. This society has published two volumes of its lectures, but they do not seen to be very important (*Mémoires de la société libre d'institution, séante au Palais national des sciences et des arts*).

The lyceums are trying to encourage social gatherings, the reading of journals and public instruction; there are three lyceums; the Republican Lyceum, the Lyceum of Arts, and the Lyceum for foreigners.

The Republican Lycéum (*Lycée républicain*) was opened on 11 frimaire (21 November 1798). I had received a written invitation from the administration to attend this ceremony. The hall and the other rooms of the *Lycée* were spacious and beautiful; the company was very numerous. One of the members of the administration announced the lectures which were to be given this winter. Millin read a fine lecture about the monuments, which had been brought from Italy to France; Lavallée gave a historical account of Charles de Wailly; Madame Pipelet read a piece of poetry before the assembly. Finally the Dumotiez brothers, who are very skilful makers of physical instruments, did some experiments with fireworks of hydrogen, which were very pretty.

The lectures begin in the second decade of frimaire or at the beginning of December and go on until 30 messidor (18 July 1799). Hassenfratz is giving lectures on technology; Coquebert on physical and economic geography; Fourcroy on chemistry; Deparcieux on physics; Sue on anatomy and physiology; Boldoni on Italian; Roberts on English; Weiss on German; Garat on history; Brongniart on natural history, and Mercier on literature. Each of these gentlemen is giving an hour's lecture twice every ten days, which makes forty-two hours or the same number of lectures in the seven months. When, therefore, a whole science, such as natural history, chemistry, &c. has to be dealt with in so short a time, the lecture has to be very short and cursory. The purpose of these lectures however is to give ladies, amateurs, and young people an idea of the wealth and use of the arts. This is also a better and more innocent occupation than wasting time on the usual material amusements. Many people, who acquire a taste for science here, are thus encouraged to apply themselves and gain a thorough knowledge. Some members have undertaken to give lectures in the monthly meetings this winter—viz: Andreux, Benoit, Celse, Deparcieux, Lacépède, Legourée, Mentelle, Millin, Perreau, and Prony.

The Lyceum of Arts (*Lycée des arts*) has elected a committee from its members, which meets on the seventh day of every decade and receives and criticizes papers, items of news, drawings, models, and other works which are sent in to the *Lycée*. Every second month a general and public meeting is held.

The work and the aim of this society can be better appreciated by a description of two public meetings. . . . I will quote some of the most important items. . . . Sobry announced through General Milet Mureau the publication of a beautiful edition of La Pérouse's journeys. In my opinion the etchings were not to be compared with the excellent etchings in Cook's last journey. . . . Colson gave a report on the antiques and monuments which are still being destroyed in the department of Puy de Dôme. In order at least to preserve the memory of those works of art which have been destroyed, Gault, the painter, has submitted some good sketches of them. Pipelot read a poem about summer, written by Madame Viot. Bonnemain described an invention of his—a fire-regulator (*régulateur de feu*) using distilling vessels. La Vieville read an allegory in verse: "The Heart and Reason" (*Le coeur et la raison*). Regnier displayed his electrifying machine with the flat conductor. He did various electrical experiments, amongst others the one which the English call High Treason. For this an image of the king is generally used, whose crown cannot be removed by anyone without receiving a violent electric shock. For this, Regnier did not use a king; he had a charged surface, one side of which was covered with tinfoil and the other was not covered On the uncovered side stood the constitution of the year II. On the covered side the constitution of the year III. The outcome was that the constitution of the second year could be touched with impunity, but if the constitution of the year III was touched, a violent electric shock was received—and that amused its advocates very much. However, should this constitution ever be changed, Regnier will certainly know how to move with the times, change the inscriptions on his surface, put the constitution of the year III in the place of that of the year II, and put the new one in the place of that of the year III, and regardless of that the surface will be just as much admired and Regnier considered a good and clever patriot. Leroi displayed the model of a mobile kitchen (*cuisine ambulante pour les armées*). Desaudray made a report on moveable and flexible wooden legs, invented by Sarnko for the use of those unfortunates who have lost their legs above or below the knee-cap. Further, he reported on a new sea-telegraph, which was simple and cheap and with which more than a million different signals can be given; it can be put together and erected in twenty-four hours. Laval, the engineer, and Montcabrier, the harbour-master, made experiments with it from Rochefort to Charente at a distance of 72,000 *toises* or nearly two geographical miles. Finally, there was music, namely an overture by Labarre, a clarinet-concerto by Dacosta, and a flute concerto by Schönscheffer. . . .

Lectures on experimental physics by Charles

In this letter I must describe to you the physical and mathematical instruments of Professor Charles. Montgolfier began with balloons, which he filled with atmospheric air, rarefied by heat. Charles, however, who was already giving highly successful lectures on physics, found a better method of filling them with the much lighter hydrogen or inflammable air. By this means his smaller sphere could raise disproportionally heavier weights and he was also the first man to rise into the air in such a machine. Naturally this gained him a great deal of fame and an extraordinary influx of students for his lectures on experimental physics, and this enabled him to acquire a collection of physical instruments. At the time of the Terror Charles lost his stipend. The *sans culottes* despised all sciences and Charles did not even earn enough to pay the rent for the large room which was required to house the valuable collection. In order to save the latter from destruction, he offered it to the Convention as national property on condition that a large room would be allotted to him in the Louvre for his collection of instruments and besides that some more rooms for him to live in free of charge. They granted him this and a few more favours, but he had to make the necessary arrangements at his own expense. The first and smaller room is for optical and astronomical instruments. There is an excellent planetarium, in which the wheels and the whole mechanism are visible, being covered only by glass. Furthermore, there are reflector telescopes, all kinds of telescopes and microscopes, two large splendid burning-mirrors, all kinds of polygonal glasses, polyhedral, cylindrical and conical mirrors, as well as good anamorphous diagrams to be used with them. Next to this room is the large room set aside for the remainder of the collection. As soon as you enter the door, you find two astronomical pendulum-clocks with compensated pendulums, which change neither in cold nor in warmth. One was made by Berthoud, the other by La Paute. On two large tables are laid out all the instruments which are used in mechanical, statistical, hydrostatic, hydraulic, and aerometric experiments, in magnetism and experiments with different kinds of gas. Near the windows just in front of the door are three large electrifying machines with splendid panes of glass almost four feet in diameter. Each of these machines has its special conductor and special battery; however they can all unite in their effect and load the combined batteries, whose coating amounts to about 100 square feet.

The instruments have all been constructed by Parisian instrument makers and have almost all the forms and shapes described and sketched by Nollet, Sigaud de la Fond, and Brisson; they look very nice and they are kept very clean and in good order. I saw here two air-pumps with tubes or cylinders of thick glass. The glass certainly does not produce verdigris from the oil, with which the piston must be lubricated, as otherwise happens with brass; twenty years of experience, however, and frequent use of the English air-pumps with brass tubes has taught me that this green oil does not attack or damage the tubes at all. Besides this, glass cylinders can easily be smashed by a blow. I consider this collection of Professor Charles the most complete collection of physical instruments, at least in Denmark, Sweden, Germany, Holland and England; I have not seen a more complete collection.

Below floor-level, in this large room, as a trophy and memento, hangs the gondola in which Charles sat when he made his first journey in the air. In this room too Professor Charles gives his lectures. On both sides a few rows of benches are placed for the audience. The lectures are divided as follows: 1. The long course on the whole of experimental physics, which will begin in November and end in March, and costs each one who attends three Louis d'or. 2. A course just on electricity. This begins in September, ends in October, and costs two Louis d'or. 3. A course on optics in the middle of summer likewise costs two Louis d'or. Professor Charles does all his experiments very well and accurately; his delivery is clear and comprehensible. I attended one of his courses on electricity, but could not agree with his explanations of various electrical phenomena. For them he used the pressure of various electrical atmospheres on one another

(*"La pression des atmosphéres électriques"*). That a body which enters the atmosphere of an electrical body or enters an electrical atmosphere, receives the opposite electric charge is an undeniable fact; we do not know the cause of this phenomenon and to call this law of nature "the pressure of the electrical atmospheres" seems an improper and quite unsuitable expression. The fee is payable in advance, but despite this, Charles always has a numerous audience and all the more so, as his lectures are the only ones of their kind in Paris. In the Central Schools and Polytechnics, of course, lectures are held on experimental physics, but these are arranged only for the young people who study there and also the collections of instruments there and the experiments set up cannot in the least be compared with Professor Charles's experiments. In all these schools I have attended the lectures as a guest and therefore I can judge them as an eyewitness. Moreover, I have not been able to find out about any other private individual whose collection of physical instruments deserves to be mentioned. Of course, some own various single instruments, such as air-pumps and electrifying-machines, but that is all.

Instrument Makers

I now come to the French instrument makers, of whom I shall mention only those who are recognized as skilful and discerning craftsmen. Lenoir is a maker of mathematical and astronomical instruments. He lives in the Rue Vendôme, where he has free lodgings in the archives of marine maps (*Dépôt des cartes de la Marine*). He has also a very large workshop and is a very clever and talented man in his speciality. Surveyor's tables, graphometers or astrolabes, box compasses, dioptric rulers, plumb-lines, levelling-instruments, surveyor's chains and rods, astronomical quadrants, complete circular scales, sectors, &c. are made by him.

He also made Borda's repeating circle (*cercle répétiteur*) for the *Collège de France* and the measurement of the degrees of the meridian, which Méchain and Delambre used. I saw in his workshop:

1. Various smaller and larger models of this repeating circle. The division was accurate and good and the rest very nice.

2. The surveyor's rod of brass and platinum, which Borda used to determine the length of the simple pendulum. It was at the same time the original from which the surveyor's rods were made to measure the base-lines at Lieusaint and Perpignan.

3. Two very fine transit-instruments which were then not yet finished. The telescope was four feet long and achromatic and the mechanism, by which the axis was placed horizontally and the axis of the telescope brought into the meridian, was very good. In the previous letters, which deal with the National Observatory, I have described this beautiful instrument in detail.

4. In his workshop Lenoir also had the wall quadrant with the five-foot radius, made by Sisson in London, for repairs. This is, of course, a different and a weaker construction than Bird's excellent wall quadrant with the six-foot radius, but all the same it is a very good instrument. The French craftsmen have not undertaken the construction of such large astronomical quadrants. For the Copenhagen Observatory here, Johann Ahl has made a wall quadrant of six feet, which is just as good as Bird's quadrants. I described and sketched it in my: *Observationes astronomicae Havnienses*, 1780–1784, Introduction.

Laroche, a member of the Board of Longitude, has free lodgings in the Louvre (*Pavillon du Midi*) and makes the best optical instruments. He showed me from his work achromatic dioptric telescopes and reflector telescopes, which were all very good. Regarding the achromatic telescopes he made the remark, that they all show the lighter colours, and particularly white, with stronger colour and stronger light, than they have in reality. He remarked further, that reflector telescopes also do not show objects quite in their natural colour, but that they lend it, according to the various composition of the metal-mirror, a yellow, red, or white light. All this is prevented by platinum mirrors, where the polish and good shine does not diminish as with the other metal-mirrors.

Laroche related that at the time of the monarchy he had worked for the king and the royal court, and at the command of the king and at his expense had done many experiments with platinum and finally discovered the correct composition and treatment, so that now he was completely master of this metal and could make mirrors from it and grind and polish them as accurately and easily as previously he had fashioned metal-mirrors of the usual composition. He showed me samples of his composition of platinum and smashed them with a hammer and the fragments were all white and fine.

Through one of Laroche's telescopes, two feet long, with platinum mirrors I looked at objects on the earth and in the sky. You can only realise clearly the wonderful effectiveness of this telescope, if you place two telescopes, one of each type, next to one another and look through them alternately. The platinum mirrors were clear and white, without yellow, red, or brown shadows or shades. The polish was excellent. Laroche poured a few drops of nitric acid on a platinum mirror, dried it after about five minutes, and even then you could not see that the acid had attacked the polish in the slightest.

It occurred to Borda that achromatic object-glasses could be improved by filling in the cavity between the crown-glass and the concave flint-glass with mastic. He spoke to me at various times of this matter as of a very important discovery. I was of the opinion, that this was really only a palliative means of concealing the imperfections of the glass. Dear Borda was somewhat obstinate in his assertions. However, as I like to be allowed my opinion, I will allow others theirs too.

These lenses filled or glued with mastic were fixed to Borda's circles and other instruments. I inquired of Delambre, Méchain, Bouvard, and other astronomers, who had used the glued lenses, how they found them. Experience, they said, had taught them that mastic lost its clarity and transparency when these telescopes were exposed to the sun's rays, and still more when the sun was observed through them, so that within a short time the instrument was completely useless. I also spoke with Laroche about this filling and just like me he was of the opinion that if crown-glass and flint-glass, of which colourless lenses consist, is well ground and the right shape, glueing and filling with mastic was not at all necessary as such an object-glass is always better than a glued one.

Dumotiez is very skilful in making physical instruments. He lives in the Rue Jardinet, not far from St. André des Arts, and makes all kinds of physical instruments, not only according to the usual French pattern, as they are sketched by Nollet, Sigaud de la Fond, and Brisson, but also according to other drawings submitted. Although the metal-work and polish cannot be compared with English work, Dumotiez's instruments are nevertheless very nice; they fulfil their function perfectly and experiments can be really well carried out with them. I have bought various instruments of his and I am very pleased with them. His prices are also very low—on the whole lower than the English ones. His nephew, a very intelligent and clever young man, helped him to manage his workshop. This young Dumotiez has had a kind of written protection from the War Minister, whereby he has been exempted from conscription until further notice. This summer, however, it was revoked and what was feared happened: young Dumotiez had to enlist and march to Switzerland. Perhaps at this moment he is dead or will have to spend the rest of his life a maimed invalid, whereby the motherland has lost a young man who could have become one of the most skilled artists in his speciality.

There is in Paris yet another maker of physical instruments—Fortin by name. He is also a very good craftsman, but requires better payment and works more slowly than Dumotiez, because his workshop is not so well staffed and equipped.

Amongst glassworkers, Betalli is outstanding. He makes barometers, thermometers, hydrometers, and of the latter the kind called in Paris Guyton's gravimeter, which the French mineralogists use, to find the specific gravity of minerals and stones. Actually it is no different from Nicholson's or Haüy's hydrometer. The temperature of the water has a great effect on it. I own such a hydrometer in which, according to the different temperatures of the water, 508 to 513 grains are required in the bowl in order to make it sink to the mark on the stem. Experience has taught me that a brass cube of 8 cubic inches in water at a temperature of 8 degrees loses 11 degrees more than in water at a temperature of 20 degrees, so that one degree of the Réaumur thermometer corresponds almost to one grain in the decrease or increase of gravity of 8 cubic inches. All this shows that in the use of the aerometer and gravimeter it is necessary to take into account the temperature of the water. It is also very difficult to decide when the mark on the stem stands exactly at the surface of the water and this can cause an error of half a grain in the determination of the specific gravity. A good hydrostatic balance is, in my opinion, the instrument by means of which the specific gravity of solid bodies as well as liquids can be most accurately determined.

Another very well-known glass-blower or barometer maker, Assier Perricat, shows his skill in glass-blowing every evening in the Palais Royal. He works very well, but his prices are higher than Betalli's. He lives at No. 231, Rue St. Antoine.

Near the Pont Neuf in the Quai de l'Horloge there are between twelve and fourteen shops, where mathematical instruments, telescopes, spectacles, burning-glasses, &c. are sold. A few such shops can be found also in the Palais Royal. There are also other shops where good optical instruments by the best English craftsmen are sold. With the English instruments it is like with all the other English goods—the law forbids them to be imported and wherever they are found they can be confiscated. In spite of this, you see them in all the shops; the ladies wear, at least in summer, only English or English-East Indian muslin; the men dress in English cashmere and cotton fabrics. This folly of preferring to use foreign manufactured goods rather than native ones prevails in France as much as in every other country.

Chapter 6

THE ARTS

It was that great patron of the Renaissance, Francis I, who in the early seventeenth century began a royal collection of paintings which was the ancestor of the later collection in the Louvre. In 1750 for the first time a collection of about a hundred paintings was opened for public exhibition. Generally, however, under the *ancien régime* royal art treasures were dispersed and unorganized. Under the *ancien régime*, art exhibitions, with the exception of the *salons*, were for aristocrats only. The Revolution brought about a democratization of art. Although there were extremists who saw art as something to be destroyed, others quickly reacted to reports of vandalism. Barère, in his report on royal buildings to the Constituent Assembly in 1791, made the suggestion that the Louvre become a museum where famous paintings should be displayed. The new function of the Louvre provides a parallel in the artistic world to what in the world of technology and science was achieved by the display of machines and instruments—as at the Conservatoire des arts et métiers.

A new source of art treasures was provided by the confiscation of church property and by the decision to transfer *objets d'art* from several royal palaces to the Louvre. A valuable collection of paintings, including Leonardo da Vinci's *La Gioconda* (Mona Lisa), was acquired from the royal palace at Versailles, but not without the resistance of the local municipality. Under the direction of a *commission temporaire des arts*, a collection of furniture, porcelain, and paintings of varying merit was assembled, thus converting the Louvre into a kind of warehouse. In the spring of 1796 the German visitor F. J. L. Meyer found one gallery of the Louvre open to the public. This was filled with a miscellaneous collection of objects ranging from scientific instruments to tables, and his impression was one of "indescribable chaos." The gallery was closed later that year.

Under the Convention there began a systematic policy of confiscation of art treasures from territory conquered by the French armies. In 1794 the first convoy of paintings from Belgium arrived in Paris, providing a magnificent representation of the Flemish school. The *Décade philosophique* described this event as the arrival of the paintings in their "true country" where they would be displayed and appreciated.[1] It was fitting that Paris should be the European metropolis of the arts.

It was left to the French armies in Italy, however, to make the most spectacular acquisition of war booty. Conquered cities were assessed a certain number of pictures as a kind of war tax. The dukes of Parma and Modena were each required to furnish twenty selected paintings. Milan and Venice also provided twenty masterpieces each, but the pope was required to supply one hundred of the best works of art in the Vatican. All the art treasures were packed with considerable (but not always sufficient) care and conveyed by ship to Marseilles. The journey to Paris was made by river and canals to avoid a rough passage by road. A "triumphal entry for the monuments of the sciences and fine arts" was planned for 28 July 1798 (just before Bugge's arrival in Paris), and a long procession of art treasures, including the bronze horses from St. Mark's, Venice, filled the Paris boulevards. After an official ceremony in which the works of art were presented to the Minister of the Interior, they were conveyed to the Louvre, where feverish activity began so that the treasures could be displayed.

[1]*Décade philosophique*, III (1794), 94.

More art treasures were brought to Paris after Bugge's visit. Sir John Carr, one of the multitude of British visitors who came to Paris at the Peace of Amiens in 1802, wrote,

> . . . I visited with a large party the celebrated museum, or palace of the arts, which I afterwards generally frequented every other day.
>
> This inestimable collection contains one thousand and thirty paintings, which are considered to be the *chefs d'oeuvre* of the great ancient masters, and is a treasury of human art and genius, unknown to the most renowned of former ages, and far surpassing every other institution of the same nature in the present times. . . .
>
> I cannot adequately describe the first impressions which were awakened upon my first entering [the gallery of the Louvre] and contemplating such a galaxy of art and genius. This room is one thousand two hundred feet long, and is lined with the finest paintings of the French, Flemish and Italian schools. . . .[2]

Many English visitors were impressed not only by the exhibits but also by the spectators. The impact of art on the lower orders was something new in Sir John Carr's experience:

> This exhibition is public three days a week, and at other times is open to students and to strangers, upon their producing their passports. On public days all descriptions of persons are here to be seen. The contemplation of such a mixture is not altogether uninteresting.
>
> The sun-browned rugged plebeian, whose mind by the influence of an unexampled political change, has been long alienated from all the noble feelings which

[2]Sir John Carr, *The Stranger in France*, London, 1803, pp. 106–108.

> religion and humanity inspire, is here seen, with his arms rudely folded over his breast, softening into pity before the struggling and sinking sufferers of a deluged world, or silently imbibing from the divine resigned countenance of the crucified Saviour, a hope of unperishable bliss beyond the grave. Who will condemn a policy by which ignorance becomes enlightened, profligacy penitent, and which, as by stealth, imparts to the relenting bosom of ferocity the subdued and social dispositions of true fraternity?[3]

The final justification of the Louvre was not, however, on the social plane but on the artistic. Its masterpieces were to be a source of inspiration to many later French painters from Delacroix to Courbet and Renoir. We must return, however, to the Revolutionary period. As a supplement to Bugge's account, an extract from the diary for 1802 of the British artist, Joseph Farington, may be quoted. This helps to convey the vital impact such a gallery could have not on the general public but on the artist himself:

> To the Painter who has to learn the principles upon which the great masters of his Art worked, the Gallery [that is, the Louvre] at Paris is the place where he may pass his hours in careful consideration and deep reflection. He may there see how Raphael thought and discriminated; how Titian by appropriate colour gave solemnity or splendour to his subject; how Correggio harmonized; and how Tintoretto attracted by his brilliance and surprised by the spirit of his execution. He will there see that the highest excellence of the respective masters was obtained by each of them following the particular bent of his mind and exerting all his powers in the way to which it was most inclined; which may cause him to weigh his own powers and consider his acquirements. This may prevent him dissipating his abilities,

[3] *Ibid.*, p. 109.

> and occasion him to confine himself to unremitting exertion to do that well of which he may judge himself most capable.[4]

Another visit Farington describes as follows:

> September 25, 1802: I passed almost the whole of this morning in the picture gallery [the Louvre] and saw for the first time the admirable whole length portrait of Cardinal Bentivoglio by Vandyke, which was brought from Genoa. It was evidently painted when Vandyke was studying the works of Titian, and it would rank with the pictures of that great master. For breadth, purity of colour, and truth of Character, I have scarcely seen it exceeded.[5]

It might be expected that a description of a visit to a great European city would include some account of its churches. It is characteristic of the period of secularization immediately following the Revolution that the typical visitor's mention of churches is limited to the expression of disgust at their desecration. From the artistic point of view the Museum of Monuments provided a repository for some statues formerly kept in churches and for specimens of stained glass. This was all the more appropriate a function for a former monastry. It constituted an important museum of medieval and Renaissance art, exclusive of paintings.

There are a few points of historiographical interest in Bugge's account. They include the Enlightenment judgment on Gothic art as barbaric (p. 184). Progress is noted in the sixteenth century (pp. 184–185), and the comment on the eighteenth century is pure self-congratulation (p. 186).

Bugge describes the libraries in Paris and particularly the Bibliothèque nationale. Frances Elizabeth King, who

[4] Joseph Farington, R. A., *The Farington Diary* (1802–1804), James Grieg, ed., London, 1923, II, 14.

[5] *Ibid.*, II, 32.

visited Paris in 1802, was particularly impressed by the National Library. She admired both the exhibits and the facilities for consulting books, which she compared favorably with the more formal arrangements for visiting the British Museum:

> ... This room [of exhibits] somewhat resembles our British Museum, with the superior advantage that it is open at all times; and every part of the library is accessible to respectable people to pass what time they please there, without expense or trouble. It is always full of the literary and curious, for whose accommodation tables, seats and writing utensils are provided; the rooms are generally fully occupied, and numbers of people in Paris pass whole days there.[6]

Finally, in this chapter on the arts we include Bugge's description of a revolutionary festival, as this may be considered an art form. What was particularly novel about such festivals was the complete involvement of the people, following Rousseau's ideas. The Roman Catholic Church had a long tradition of festival days with appropriate celebrations. The new civic and secular festivals were to replace the Christian celebrations as the republican calendar was to replace the Gregorian calendar. It was realized that emotional celebration was particularly important for the illiterate masses. The painter Jacques-Louis David was largely responsible for the organization of the new festivals, which were planned on a large scale and with a view to the maximum political propaganda. The apotheosis of Voltaire in July 1791, when his body was solemnly brought to the Panthéon, provided something of a precedent. Most famous of the festivals was probably that of the Supreme Being on 20 prairial year II (8 June 1794), which marked the climax of Jacobin power. Under the Directory, public ceremonies designed to encourage

[6] [Frances Elizabeth King], *A Tour of France*. 1802, 2nd ed., London, 1814, p. 34.

feelings of fraternity and patriotism continued to be used as an instrument of popular education. A particularly important festival was that marking the anniversary of the Republic, and Bugge describes the festival held on the first day of the Republican year VII (22 September 1798).

Near the Louvre, or National Palace for Arts and Sciences, is a building appropriated to collections in the fine arts of drawing, painting, and sculpture, under the name of the Central Museum of Arts. All the foreign pieces of art formerly seen in France, together with the paintings and statues which have been since acquired from Belgium, Lombardy, Venice, Rome, and other States, either by the force of conquest, or by express conditions in treaties of alliance and neutrality, are now formed into one general collection. The large and strong four-wheeled carriages which brought those subjects of art from Italy, are now standing in the garden of the Louvre.

The entrance to the Museum of Arts is from a large square in front of the Louvre, and close by a corner formed by this square and that palace. All the way in entering, statues of bronze, and busts of marble, are presented to view in porticoes. In the front room, among other pieces of sculpture, are four beautiful colossal slaves, which once stood by the pedestal of the statue of Louis XV in the Place des Victoires. Below the entrance into the stair-case are several statues, which have been brought from Italy, and on the different landing places of these noble stairs, are various fine models of gypsum.[7] On entering the second floor there is a large front room, or salon, with sky-lights at top. On the left hand is the gallery of Apollo, containing only sketches and some crayon paintings, most of which were brought from Holland, Belgium, and Italy. On the right hand, is an excellent and singularly extensive picture gallery. It is a room of no less than five hundred feet in length, and was formerly filled with paintings, small statues, busts, idols, vases, mechanical contrivances, mathematical and philosophical instruments, models

[7]"Gibs"—J. J.

of buildings; and, in short, it was that kind of disorderly jumble which some virtuosos are fond of amassing, in what they call a cabinet of curiosities. This great gallery is at present undergoing an alteration and new arrangement, on which account it has been shut up for the whole summer. The managers of this museum announced on 18 brumaire year VII (8 November 1798) "that they had made considerable progress in arranging and putting into proper points of view, the paintings produced by the Flemish and French Schools, in a part of the great Gallery, which they intend opening as soon as possible; that they will then publish a new catalogue or explanation of the paintings in the gallery of Apollo, and will work with all possible dispatch, in preparing the place where the Italian statues are to be exhibited to public view."

The Italian paintings have been publicly explained in two catalogues. The first of which extended from 18 pluviôse to 30 prairial year VI (from 6 February to 18 June 1798), and included the pieces brought from Lombardy, that is, from Parma, Piacenza, Milan, Cremona, Modena, Cento, and Bologna. To this catalogue have since been added, some Italian pieces from Versailles, in order to collect the whole Italian school into one point of view.

The managers have had the candour to acknowledge, that some of those masterpieces of art are in such bad condition that they cannot be exhibited. This seems tantamount to a confession, that they have been much injured on the journey, if not totally abraded and destroyed. In particular, it is known, that an excellent portrait of Raphael, by Foligno; the Holy Virgin and some Saints, by Bellini; the repast at the house of Levi, by Paul Veronese; the Marriage of Cana, by the same master; St. Peter, the martyr, by Titian; and several of the statues brought from Italy, have suffered greatly from the length of the journey.

[Bugge then lists a total of 141 paintings by a variety of artists (mainly Italian) including Bellini, Barbieri, Leonardo, Tintoretto, and Breughel, together with the place of origin of each.]

Out of these one hundred and forty-one paintings, there were thirty-six formerly in France; this reduces the number brought hither from Italy to one hundred and five. They will all be hung up in the Great Gallery, as soon as it is ready for their reception.

The other catalogue or explanation consists of pieces brought from Venice, Verona, Mantua, Pesaro, Fano, Loreto, and Rome. . . . This second collection I have visited more than once. I went there the first time possessed with the idea of seeing something great and beautiful in the enchanting art of painting; but I must say that the works of Raphael, Guido Reni, Paul Veronese, Andreas Sacchi, and other great masters, exceeded my most ardent expectations. Yet I cannot forbear mentioning, that I also saw some pieces which were not at all pleasing to me: but they all exhibit authentic traits of the times, and ought to be there, as they form a history of the art, its progress and perfection in design, colour, light, shade, &c.

With respect to the statues brought from Italy, a plan has been drawn of the other in which they are to be set up, in a number of adjoining rooms, which are to be prepared and embellished. In the middle of each, is to be erected a large statue of superior beauty, such as Laocoön, the Farnesian Hercules, the Apollo Belvedere, &c., the rooms to derive their names from these statues, and to be called the Salon of Laocoön, the Salon of Hercules, &c. &c. The statues of less size and beauty are to be set up in those salons.

In the Great Salon, or Gallery of the Central Museum, where the Italian paintings just noticed are hung up, are annually exhibited the performances of French artists now living, together with those of their pupils. Such a collection was exhibited in the year VI, for four weeks, commencing on 1 thermidor (19 July 1798). It consisted of four hundred and forty-two pieces, some drawn in oil colours, and some with Indian ink. A collection thus extensive, and executed by so many different young artists, must necessarily possess different degrees of merit. . . .

Among the portraits done with oil colours, and as large as life, that of Professor Charles seemed to me to be very happily designed, and well painted. He is drawn in a kind of grey silk morning gown, in which, I am told, he used to lecture on electricity. In his hand is a solar microscope, which is a very proper emblem, as he had a remarkably fine apparatus for optical experiments, and his lectures on that science, which he delivered in the summer season, were particularly admired.

The statuary performances consist of forty-nine pieces. Among them is a bust of the worthy Daubenton, at the age of eighty-three. There are eleven architectural drawings, and twenty-six copper-plate prints, among which is a fine portrait of General Marceau, engraved by his brother-in-law Sergent. This piece is extremely well coloured.

I have to mention in this place, that there is at this time prepared, at the uninhabited palace of Versailles, "A General Museum for the Paintings of the French School." This museum occupies eight large apartments, on the upper floor. The paintings have been all taken from cloisters, churches, and collections belonging to the emigrants, and to the former government. The whole is well arranged, and has a very good effect. It contains a number of excellent paintings; but there are also some which have but a moderate appearance, when examined after one has seen the great master-pieces of painting which have been brought thither from Italy.

The little Augustinian[8] monastery at Meudon, now shut up, contains a collection of French monuments. The decree of the National Convention to abolish every vestige of royalty, or anything that might recall the days of feodality, was considered as the signal for desolation, plunder, and rapine over all the kingdom of France. The statues of kings and others without distinction were hewed down and levelled with the dust. The noblest and happiest efforts of the pencil were rent in pieces, and scattered in the air. Entire series of the most precious medals, the labour and research of ages, were stolen or consigned to the crucible. All the monuments and epitaphs within reach of the hand of fury, were broken to pieces. The greatest part of the labours of the first artists, collected in different parts of the

[8]"Augustine"—J. J.

world, shared the same fate. The vandals and barbarians, who rather resembled the furies let loose from hell, than human beings, vented their ungovernable rage on the choicest productions of taste and genius. In the Convention and Revolutionary tribunals, the most profligate and abandoned boasted of the revenge which they took on the arts. The enlightened Grégoire ventured at length, at the risk of his life (lest he should be accused of being attached to the old system) to stand forward as the advocate of the Muses. On 31 August 1793, he addressed a letter to the Convention, in which he painted in the most natural and lively colours, the eruption of this vandalic horde into the sanctuaries of science, and the excesses which they committed on monuments that lent immortality to mortals. This eloquent epistle at first had little effect: at length, however, the Convention began to think of converting the public monuments to national property. For this purpose they ordered them to be collected and deposited in the small Augustinian cloister. . . .

Lenoir [was appointed inspector of the Museum and he] began without delay to arrange and repair the mutilated statues, &c. with unremitting industry, at as little expense to the public as possible. The general plan of classification is to arrange the statues, &c. in centuries in rooms decorated in the taste of each age. Three rooms are already devoted to this purpose, viz. the thirteenth, sixteenth, and seventeenth centuries. These rooms are extremely neat, some of the statues are raised on pedestals, and others placed against the wall. The monuments thus arranged, and erected in the three rooms, amount to upwards of two hundred.

I shall now give a short description of these monuments. The first collection is the Grecian antiques, twelve tomb-stones of fine marble, with Greek inscriptions and bas-reliefs, some statues taken from Richelieu's garden, and amongst the rest a highly finished Bacchus, as large as life, with his thyrfis in one hand, and a bunch of grapes in the other, and Meleaeger in the chase. They are fine statues of Parian marble. Those antiques in all amount to twenty-six in number. Of Celtic monuments there are four altars. . . .

The room set apart for the monuments of the thirteenth century is already completed. The dome is vaulted in the Gothic style, with a blue ground, studded with gilt stars, the sharp pointed bows or arches support each other, ornamented with roses according to the taste of the day. . . . The rudeness of the age is visible in all these monuments; some indeed evince a greater progress in taste and execution than could be expected in those times.

In the room sacred to the fourteenth century, there are thirty-eight monuments, most of which were erected in St. Denis to the memories of the Kings of France, such as Louis X, Philip V, Philip of Valois, Charles V. The figures are almost all of marble. . . .

In the room for the fifteenth century there are fifteen monuments, consisting of Queens, Princes, Princesses, &c. taken from the royal cemeteries in St. Denis. . . .

The room designed for the sixteenth century is finished with great taste. Two academic figures, executed by Barthélemy Prieur, are placed over the door. The joints of the door are of yellow streaked marble. The ceiling is ornamented with arabesque, according to the taste of the times. This hall contains fifty-three monuments. The monument of Louis evinces that the arts already began to make a rapid progress towards perfection. The King's statue, as well as that of the Queen (Anne), are of excellent workmanship. The twelve Apostles are arranged in twelve richly ornamented arcades, in which the artist has exhibited considerable taste in the style and representation. In the four corners there are the four cardinal virtues in a natural size, the whole rests on a pediment of black marble, on whose edges there are bas-reliefs, representing the victories of Louis XII. This fine historical monument has suffered much from jacobinic rage. The heads, nose, arms, and hands, are broken from the figures. . . .

The monument of Francis I is erected in a particular chapel, set apart for that purpose. The King and his Queen Claudia are represented by two marble figures, somewhat larger than the natural size, extended as dead. The artist may be said to have disputed the prize with nature in the execution of this monument. The privation of life in the countenance and muscles is finely expressed. The pedestal on which these figures lie is adorned with a bas-relief, representing the victories of Francis, different genii with extinguished torches, allegoric figures, &c. The roof is supported by sixteen fluted columns. Francis and his Queen are grouped on the ceiling in their gala robes, together with the two Princes, their sons, and their daughters, all of fine marble. The whole was designed by Primaticcio and sculptured by Jean Goujon. This costly monument, the first perhaps in France for design and execution, formerly stood in the church of St. Denis. It suffered very much in the paroxysm of popular tumult; but Lenoir has been very successful in re-instating the fragments, according to the original design, which he had copied in his youth, whilst a student in works of this kind. . . .

The mausoleum of Cardinal Richelieu was designed by Lebrun, and executed by Girardon, and is considered as his master-piece. The pedestal is fourteen feet long, and five feet nine inches broad. The Cardinal's figure is six feet high, placed betwixt two female figures, one representing Religion, the other History, with two Genii, each two feet and a half high. This mausoleum did not escape in the war that was waged against the productions of genius. Under the medallion of Descartes, suspended to a pyramid of black marble, there are two inscriptions, one in Latin and the other in French. . . .

There are two very neat monuments raised to the memory of Colbert and Louvois, the two great ornaments and support of the reign of Louis XIV. These monuments were executed by Girardon, Coyzevox, Tubi, and Desjardins. . . .

In addition to these, there are thirteen statues of Christ, and the Holy Family, of an extraordinary size, that were all collected from the pillaged churches in Paris, with sixteen bas-reliefs, some in marble and some in bronze.

Of busts, a great number has been saved and erected, such as those of Henry III, Louis XIV, and Louis XV. Of statesmen and warriors there are fifteen, and amongst others those of Sully, Mazarin, Richelieu, Colbert, Turenne and the great Condé, &c. Of learned men, Peyrese, Regis, Quinault, La Fontaine, Molière, Corneille, Racine, and Boileau. Of artists, Mignard,

Poussin, and Lesueur, painters; Le Nôtre and Mansart, architects and Sarazin and Puget, the sculptors. A few of these busts are of bronze, and the rest marble. I also observed a small piece of Mozaic work, finely executed, representing Saint Hieronymous in the desert.

There are thirteen monuments already collected, the works of the eighteenth century, excelled by none in correctness of design, and felicity of execution. . . .

The busts and medallions consist of statesmen and generals; the Regent Philip of Orleans, Marshal Asfeld, Count of Saxony, D'Argenson, and Montesquieu. In the learned class are Destouches, Fontenelle, Astruc, Helvetius, Piron, Belloy, Voltaire, J. J. Rousseau, Buffon, Diderot, Gluck, Raynal, Bailly, the famous astronomer and historian, the first Mayor of Paris, Vaucanson, &c. . . .

I shall now give you a short account of the stained glass. In the depot of the thirteenth century, there are three large church windows, with panes of painted glass, the work of that age. They were taken from the abbey of St. Germain-des-Prés, and represent moral subjects, particularly those of domestic life. . . .

The rooms, which are reduced to order, namely, those of the thirteenth, sixteenth, and seventeenth centuries, are very neat and commodious; the fourteenth and eighteenth, as yet lie neglected, unless we except those that are erected in the court, gardens, passages, and rooms of the aforesaid Augustinian cloister. This promiscuous heap of warriors, statesmen, knights, men of letters, saints, prelates, monks, and nuns, creates an agreeable surprise at first view but the contemplative spectator finds himself immediately disposed to ask this question, What right has the present generation to destroy those monuments which their forefathers erected to perpetuate the memory of their relatives or friends? Have they not afforded maintenance and support to many artists, and fanned the sparks of genius and emulation? What must the living artists think when they see the precious works of their masters exposed to the caprice of a licentious mob? Have they not serious cause to be alarmed for the future fate of their own labours, and that a single day may destroy the labours of ages? What must be their feelings when they enter any one of the churches which are still open in Paris, stripped of their ornaments, the naked walls disfigured with holes, and the floors on which those monuments stood covered with dirt and gravel; what an awful sensation, when Reason has resumed her throne, to behold the trophies of the arts thus scattered and annihilated!

Bibliothèque nationale

The National Library, formerly the King's Library, is situated in Rue de la Loi, formerly Rue Richelieu, opposite to the great Opera House. The south side faces Rue Neuve des petits Champs, and its north side is in the Rue Colbert. The building of the library, with its appurtenances, is very large; its length in Rue de la Loi being no less than eighty-five *toises*, and its breadth between the two streets above mentioned, twenty *toises*. In the court of the National Library, is a fine statue of bronze, representing a woman standing on one foot, in a very easy and natural attitude. The principal floor of the building, which surrounds this large court, is entirely filled with books, from the floor to the ceiling; it is surrounded by a slight gallery, from which one can reach the books on the upper shelves. At the windows, and in different parts of one of the wings, tables have been placed for the accommodation of readers. While the weather continued mild and fair, I always found from forty to sixty persons, some of them ladies, reading at those tables. The library is open every day, except the decade days, from ten to two, for the accommodation of readers; but no books are lent out. For such as only wish to see the library, it is open from ten to two, every third, sixth, and ninth day of the decade.

In a small recess of one of the four sides of the library, is a group of about five feet in height and six in breadth, erected in the time of Louis XIV. It represents, as far as I could collect, Parnassus with Apollo and the Muses, and several attributes applicable to the era of that Monarch. There are also in the library some busts of celebrated French literati, and of others, who have contributed to the improvement and augmentation of the library.

In the other wing of the library, a very large perforation in the floor presents two large globes, the celestial and the terrestrial, which stand on the floor below, and their upper parts project above the floor of the library. These globes are thirty feet in diameter, . . . and are the largest I have ever seen. . . .

Capperonnier, the present librarian, supposes the library to contain about 300,000 volumes. It is very incomplete in modern literature; for, since the year 1789, no new books have been added to it, not even French, and much less foreign productions. Of this last description, several capital works seem wanting; so that in the midst of this great opulence, a kind of literary penury is still felt. The national and other libraries have received considerable augmentations from the libraries of monasteries and emigrants. This is an easy, and a very cheap method of increasing a stock of books.

The manuscripts, to the number of 80,000, are in more retired apartments. The oriental manuscripts are kept by Langlès; those in Greek and Latin by Laporte Dutheil; and those in modern languages by Legrand. The manuscripts are divided agreeably to this classification, and are well arranged. Since these subjects are foreign to my sphere of study, I shall only relate such observations as I made, in a cursory manner. Here is a complete collection of Colbert's letters in about sixty volumes. A volume of letters, some in English and others in French, written by Henry VIII of England, in a good, legible hand. A volume of letters from King Henry IV of France to one of his mistresses: his handwriting is tolerably neat and legible, and he has expressed himself with much vivacity and gallantry. To indicate the ardour of his amorous attachment, he says, "*Je vous aime plus, que vous aimez vous même*" (I love you more than you love yourself). Here is a large collection of the French King's *heures*, or missals, all written very beautifully on the finest vellum, and embellished with elegant borders and fine drawings, most of them scripture histories. On every leaf of one of those missals, is a beautiful drawing of a flower, with its name in Latin and French, so that it forms a collection of botany as well as religion. Vanquished Italy has been obliged to contribute her mite to the treasury of the National Library; for all the most valuable printed books and the scarcest manuscripts, have been taken from the Italian libraries. Among those Italian manuscripts, I particularly observed two Codices in parchment, a Terence, and a Horace, from the library of the Vatican. I am no hunter after various readings; yet it is possible these Codices have no critical merit, but are remarkable only for their external beauty and excellent preservation.

Two rooms belonging to the library are filled with a large collection of prints, which are under the superintendence of Joly. Some pieces are hung to the walls, but most of them are in portfolios and cases. Here in particular is a collection of about sixty volumes of prints of remarkable transactions and events, in the history of France, arranged according to the year, or reign, down to the time of Louis XV.

The collection of antiques and coins is at the end of the library: the keepers are Barthélemy and Millin. The latter gentleman is remarkably attentive to strangers, and every seventh day of the decade has an agreeable party to drink tea at his house, where he is glad to see foreign travellers. . . .

Millin reads public lectures on archaeology every second, fifth, and eighth day of the decade. He is editor of the *Magazin Encyclopédique*, and is well known by his other publications. The collection of antiques and coins is not open to the public, but is to be seen by particular permission. Millin has the goodness to shew this collection to . . . myself and other travellers.

Straight against the entrance and over the chimney-piece, various Egyptian antiquities meet the eye; such as an altar of basalt, an Isis, Anubis, and several curiosities of bronze, stone, and burnt clay. Here is a mummy taken to pieces, the upper covering having been taken off, and extended upon the wall: it is remarkable for its fine colour and drawings, which without doubt were emblems of religious ceremonies. Near the fire-place are drawers, containing French and other medals, chiefly of gold. On the wall to the right of the door, are hung up lamps, and sacrificing knives, and vessels of bronze. Between the windows, on the same side, are several large chests with glass-lids, containing antiquities of the primitive times of Christianity, chiefly Greek. In the third and fourth divisions, are stones cut in bas-relief, some of them Greek, and others Roman productions. Almost all of them have been executed in stones, which have their laminae of different colours, disposed in such a manner, that the features of the figures had one colour, but the hair, helmet, clothes, &c. different ones. There are among them many beautiful and excellent pieces. At one end of the room are several warlike instruments of different Indian nations. On the floor, stands a large antique marble table, with a Latin inscription; and the walls are decorated with the shields of Scipio and Hannibal, which were once suspended in temples. They are of silver, and of very beautiful workmanship. By the side of them are placed the arms of Francis I, such as his helmet, shield, sword, battle-axe, and spurs, all of steel, inlaid with gold, and most exquisitely formed. On the shield are Arabian drawings, executed with regularity and taste. His stirrups which are of silver, gilt and carved with open work, are placed underneath. . . .

After having viewed this room, where everything was arranged in the best manner, Millin conducted us to the third floor, where are two apartments, which contain a very large and remarkable assemblage of antiquities; Etrurian vases of extraordinary magnitude; a bathing vessel of porphyry in good preservation; figures in bronze; sacrificing knives, lamps, household furniture, &c. not suspended separately on the walls, but placed here and there along the floor, as convenience admitted. Those apartments on the third floor, seemed to be more distinguished as antiquarian lumber rooms, than for any regular arrangement of the many valuable curiosities which they contain. Millin, for above three years, has been requesting money for constructing cases and shelves for arranging and containing this chaos of antiquities; but his applications have not yet been attended to. He is full of zeal and activity in this his favourite pursuit: he complains that the study of ancient literature and arts are not only neglected, but totally despised, as unnecessary for forming a good taste and accurate ideas of the fine arts. On the first of vendémiaire, or last of September, none of the attendants belonging to this collection had received any salary for the preceding eight months.

By a decree of 10 germinal year III, a school was instituted, adjoining to the National Library, for the modern oriental languages, where public lectures are delivered by Langlès, on the Persian and Malay languages; by Silvestre de Sacy, on the common and learned Arabic; and by Behenam, on the Turkish and Tartarian.

While I was in Paris, my countrymen, Dr. Müller, Dr. Englestoft, and Dr. Thorlacius were also in that city. As those gentlemen regularly visited the National Library, they will be able to give very particular accounts of that establishment; nor is it to be doubted, that when opportunity serves, they will favour the public with some of their observations.

Besides the public libraries mentioned in this and preceding letters, there are the two following:

1. The library of the Arsenal, which is supposed to contain 75,000 printed volumes, and 6,000 manuscripts, and which formerly belonged to the Count d'Artois. It is open every first, sixth, and eight day of the decade, from ten till two.
2. The library of the Pantheon, formerly the library of St. Geneviève, which consists of about 100,000 printed volumes, and 2,000 manuscripts, and is decorated with different marble busts of French *literati.*

Before I quit this subject, I must remark, that on the first appearance of terrorism, the libraries, paintings, natural curiosities, and instruments of such as were banished or put to death, were partly destroyed, and partly carried off. But the more prudent put a stop, as soon as possible, to those robberies; and it was resolved, that all such articles should be considered as national property, and be collected and preserved, until further orders. Of such collections of books, three depots have been formed: one in Rue des Capucins, one in Rue des Cordeliers, near the Medical School, and one near the Central School in Faubourg St. Antoine, *ou ci-devant Jésuites.* Those books are now arranged and distributed among the libraries of other Institutes in Paris, and in the departments; and I have often seen cart loads of books taken from those collections. . . .

The celebrated and indefatigable Millin, in the commencement of the year 1792, published a collection of monuments in four volumes. He had been at great pains and expense in travelling over France, to collect the most remarkable either for design or execution that the kingdom could boast. They were copied and engraved with great accuracy. He had also copied a great number of epitaphs and inscriptions, which he illustrated with many historical remarks, so as to render them very interesting to the historian and antiquary. In the month of November 1798 he published the fifth volume under the title of *Antiquités Nationales.*

Millin's house is the resort of all men of genius and taste. It is the only one in Paris where a traveller can form an immediate acquaintance with Frenchmen and strangers. Every seventh day in each decade he gives what is called his literary tea. The company begin to assemble about eight or nine o'clock. The table in the first room is covered with French and foreign journals, and new publications of merit. The inner chamber is occupied by ladies, who play on the *claveçin,* and accompany it with their voice, which has a pleasing effect, especially in filling up the pauses of conversation. About eleven the company is treated with tea, punch, and cakes, his good mother pays the utmost attention to the guests, and seems highly gratified in rendering them every courtesy in her power. About twelve they all retire.

In those circles I have found thirty and forty persons at a time. M. Millin has a book, in which every traveller writes down his name to enable him to preserve the remembrance of each. I am glad of this opportunity to return Mr. Manthey, the Danish secretary of legation, my sincere thanks for having introduced me to M. Millin, in whose house I have passed many agreeable evenings, and where I had frequent opportunities of forming many valuable acquaintances, which rendered my stay in Paris at once amusing and instructive.

Revolutionary festivals

On the fifth complementary day [21 September], about eight o'clock in the evening, there was a general discharge of artillery and at nine another from the cannon planted before the palace of the Directory, and along the banks of the Seine. This discharge was immediately followed by six hundred rockets from the Pont-neuf, which ascended to a considerable height, and formed a beautiful appearance in the air. The public offices and telegraphs were hung with lamps, lighted up with different colours, which had a very pleasing effect, as they were arranged to the best advantage. Glass lamps are not used in those illuminations, but flat lamps of potter's earth from three to four inches. They are not filled with oil but a substance prepared from the offals of oxen, calves, and lambs, which are purchased in the slaughter-houses for that purpose. In serene weather they burn very clear, but wind or rain immediately extinguishes them.

22 September 1798 was a peculiar festival. There were rowing matches on the Seine, and wrestling in the Champ de Mars, for small prizes, such as ribbons, &c. The victors were immediately invested with the prize, and sometimes carried off on the shoulders of the populace in triumph, particularly if the contest was long doubtful. In this lift I saw fourteen or fifteen young men, well formed by nature for such athletic exercises.

This amusement was succeeded by the entrance of two chariots. Some of the victors of 14 July stood upright in one, and some of these of 14 August in the other. A party with lighted brands set fire to two figures of wood, the one representing Despotism, and the other Fanaticism, and then danced round the blaze.

In the afternoon the Directory, Ministers, &c. assembled in the Military School, from whence they moved in procession. . . . A group dressed in the costume of the ancient Gauls walked before the Directory, with a banner containing the names of all the departments. The following lines were written on the back of this fane:

La République les a tous réunis,
Ce n'est plus qu'un même peuple.

A trophy was borne on one side of the departmental ensign,

formed of the shields of the Batavian, Cisalpine, Helvetic, and Roman Republics, with the following inscription:

Que leur alliance avec le peuple François soit éternelle.

As soon as the procession had reached the altar, raised to the genius of the country, the departmental ensign, and the trophy of the allied republics, were placed at the foot of it, with great ceremony, accompanied by a triumphal song. Treilhard, the president of the Directory, delivered a speech, in which he congratulated the French nation of the success of their arms, and the prospect of returning peace. An ode was then sung, composed for the occasion, the words by Chenier, and the music by Martin.

The president then read over the names of the citizens who had contributed to the stability or happiness of the Republic in the course of the preceding year, either by their personal bravery, patriotic essays, inventions, industry, &c. . . . Horse and chariot racing followed. . . . The government seems to know the Parisians well, and how easily they may be managed with spectacles of this kind, like the Romans, who only wished for bread and shows (*panem et cirences*). . . .

Chapter 7

THE METRIC SYSTEM

As the purpose of Bugge's visit was to represent his country in the establishment of the metric system, it is appropriate to include some account of the new system of measurement (despite its complete omission from the English translation of 1801) and to comment on the part played by the international commission. Bugge's original account is a very long and detailed one. Only short extracts are provided here, but there is a proportionately longer introduction.

Already in 1790 an attempt had been made to obtain some international authority for a new system of measurement by collaboration with England. The British Parliament had discussed the question of uniform measurement within Britain in 1790, and this had encouraged the French National Assembly to think of collaboration. Its decree of 8 May 1790 asked Louis XVI to approach the British government with a view to adopting the same new unit of measurement. If the two most powerful nations in Europe had agreed on a new unit, the rest of Europe would soon have followed. The United States, too, in the person of Thomas Jefferson, ambassador to France, was interested in the question of unification of measurement. In France the task of selecting units of measurement was passed on to the Académie royale des sciences. The committee appointed by the Academy evidently preferred to make decisions based on the authority of the Academy rather than depend on diplomatic negotiations in an increasingly difficult political situation. Its report of 19 March 1791 stated, "We have not thought it necessary to wait for the collaboration of other nations, either for the choice of units of measurement nor to begin our actual work."

The initiative having been taken by the Académie des sciences, the earlier ideas of international collaboration fell into abeyance. Yet the problem of international acceptance was not forgotten by those who tried to frame a new system. They therefore sought a "natural" unit. The length of a seconds pendulum was at one time considered a suitable standard. As, however, this varies according to latitude, the latter would have to be specified. A convenient latitude for a European standard would have been 45 degrees but because this line passes through Bordeaux it was open to the charge of being a French rather than an international unit. Finally it was decided to take as the "natural" unit a fraction of the distance between the equator and the North Pole. The meter was introduced as a length equivalent to one ten-millionth part of this distance.

It was obviously impossible to measure the total distance between the equator and the Pole *directly*. The principle used was to measure with great precision the distance between two points on a meridian in Europe, this distance constituting a known fraction of the whole arc. Irregularities on the earth's surface, however, prevent even the direct measurement of long distances on land and the method used was one of triangulation from a series of base points, as in any smaller surveying problem. The calculation of the fraction of arc depended on the exact determination of latitude and was, therefore, an astronomical problem. Ordinary practical difficulties made the measurement of the meridian from Dunkirk to Barcelona a work of years rather than months, but the delay was aggravated by the abolition of the Academy (8 August 1793) and then the suspension of the temporary Commission of Weights and Measures which had been allowed to continue this important aspect of the work of the former Academy.

Thus it was only in 1798 that France was ready to go from the provisional meter (a working standard related to earlier and obviously approximate measurements of the meridian) to the definitive meter. On 20 January 1798 the question of international collaboration was again raised—this time at a meeting of the First Class of the Institute. The official body of French science resolved to ask the French government to invite foreign governments to send scientific representatives to Paris to take part in the final stages of the establishment of the metric system. Talleyrand, as Minister of Foreign Affairs, sent an invitation, not to all European countries but to friendly powers and neutral states, to send scientific representatives to be informed of the work done by the French on the metric system so far, and to take part in the final and formal adoption of the new standards.

The delegates had been invited to arrive in Paris for the beginning of the Republican year VII (about 22 September 1798), but some arrived late (the Spaniards) whereas others arrived in advance of the date specified, Bugge, for example, arriving on 18 August. The foreign scientists who eventually came to Paris were the following:

Batavian Republic	Van Swinden and Aeneae
Cisalpine Republic	Mascheroni
Denmark	Bugge
Spain	Ciscar and Pédrayès
Helvetian Republic	Trallès
Ligurian Republic	Multedo
Kingdom of Sardinia	Balbo
(replaced by provisional government of Piedmont)	(Vassalli Eandi)
Roman Republic	Franchini
Tuscany	Fabbroni

It will be seen that, translated into terms of modern European geography, Italy was particularly well represented. The inclusion of Denmark, the Netherlands, Spain, and Switzerland meant that virtually all of western Europe was represented except Great Britain, Portugal, and Sweden.

The official committee on the metric system had quite a long and tragic history. A temporary committee of 1793 was "purified" by a decree of the Committee of Public Safety of 23 December 1793 to exclude several prominent members, including Lavoisier and Delambre. By the Republican year VII (1798–1799) the membership of the committee was Borda, Brisson, Coulomb, Darcet, Haüy, Lagrange, Laplace, Lefèvre-Gineau, Méchain, and Prony. Monge and Berthollet had previously been prominent members but they had been replaced after their departure for Egypt with Bonaparte.

At the beginning of the joint meetings in the autumn of 1798, the foreign delegates were shown the instruments used in the previous determinations. Three joint committees of the French and foreign men of science were then appointed to consider different aspects of the metric system:

1. A committee to compare the measures used with different specimens of the old unit of length, the *toise*, with regard to temperature variation.

2. A committee to redetermine the exact length of the meridian passing through Dunkirk in the north and Montjouy in the south of France, as well as the redetermination of latitudes. The tedious and complex calculation of triangles was carried out independently (but using the same data) by Van Swinden, Trallès, Delambre, and Legendre. The redetermination of latitudes involved nightly vigils by Méchain in the Paris Observatory, which resulted in a constant value and made the other members of the committee impatient because of the delay.

3. A committee concerned with the determination of the unit of weight. They finally took as the unit (one kilogram) the weight of a cubic decimeter of distilled water taken at the maximum density and weighed *in vacuo*, which was 18,827.15 grains according to the standard, the *poids du marc* of Charlemagne.

The reason for the invitation to the foreign delegates was undoubtedly to give the new metric system greater authority by having some international support. It was also good propaganda for the French state. If these events had taken place under the Consulate rather than the Directory, there is little doubt that Bonaparte would have exploited them even more. Nevertheless the conclusion of the work of the joint committees was marked by public ceremonies on 22 June 1799 in which the prototype standards of the meter and kilogram were presented successively to the two legislative bodies, the Council of Ancients and the Council of Five Hundred. The speech made on this occasion by an unnamed orator (possibly Laplace) emphasized the international character of the work. This aspect was taken up in the reply by the president of the Council of Ancients, Baudin, who said that he looked forward to the breaking down of national frontiers:

> We already have an indication of this in the co-operation of so many learned collaborators from allied or neutral nations who have been associated with your work. If the form of the governments which have sent them is not the same,[1] they are all citizens of that republic of letters which is essentially fertile in generous sentiments. . . .[2]

The circumstances in which the foreign delegates had been invited to Paris may have led some to sense a feeling of paternalism if not actual condescension. This was not the impression which the Institute or the government wished to convey, and at the ceremony at which the standards were presented the members of the commission were named in strict alphabetical order without distinction

[1]Although most of the governments represented were republican, Spain and Denmark were still ruled by kings.

[2]*Base du système métrique décimal. . . . Exécutée en 1792 et années suivantes par MM. Méchain et Delambre*, Paris, 1806–1810, III, 653.

by nationality, thus demonstrating the principle of fraternity. An account was to be given to the Institute of the work of the joint committees and the choice fell deliberately on two of the visiting men of science, the Swiss Trallès and the Dutchman Van Swinden. Although these arrangements suggest some stage management, they also reflect a genuine desire for international collaboration. That this work was carried on in the midst of the Revolutionary wars obviously limited the extent of such cooperation.

The French authorities had stated that the meetings would take place at the latest by 5 October 1798, but a delay was caused by the fact that Méchain and Delambre were then still carrying out a redetermination of the base line. On 26 November the waiting commissioners were told that Méchain and Delambre had returned to Paris and two days later the first joint meeting was held at the Dépôt de la Marine. Bugge attended further meetings on 1, 5, 11, 22, 28 December 1798 and on 3 January 1799, some being held at the Institute. His patience was then exhausted. The Danish astronomer wrote to the Minister of the Interior, reminding him of the delays and pointing out that the definitive standards for the meter and kilogram had still not been obtained, nor was this likely before the summer. Having several duties in Denmark, Bugge said that he felt obliged to return home and he left Paris in February. The French government had presented him with copies of several recent French scientific books and also a collection of maps; the Institute too had been particularly hospitable. There was, therefore, some resentment (suggested by an article in the *Décade philosophique*) at his criticisms and his decision to leave early. Although Bugge had left France by the time the standards had finally been settled, he continued to keep in touch with his former colleagues at the Institute and he calculated the relationship of the kilogram and meter to Danish weights and measures.

The establishment of the metric system had various

beneficial effects on scientific research quite apart from its direct utility. Thus, as Bugge tells us, the original choice of zero degrees centigrade for the temperature of the water in the determination of the standard kilogram was modified when it was reported that water had a maximum density at about four degrees centigrade. Experiments were carried out to confirm this important datum. The accurate measurement of standards of length necessitated a further investigation of the coefficient of expansion of various metals. Again, the use of platinum for standards of length led to significant research on the chemical and physical properties of this metal. Instrument makers were called on to display their utmost skill and ingenuity. Lenoir, for example, constructed a massive copper rule fitted with a series of Vernier scales (*comparateur de Borda et Lenoir*) used for the comparison of standards of length, and ultrasensitive balances were constructed by Fortin.

In one case the availability and accuracy of a particular instrument seems to have influenced the method used. This was Borda's repeating circle (*cercle répétiteur, cercle de réflexion*). The idea of this came from the German astronomer Tobias Mayer, who in about 1752 thought of substituting a complete graduated circle for the usual octant or sextant used to measure the elevation of a star. The purpose was to achieve speedier and more accurate readings by taking the mean of a number of consecutive readings without returning to the zero. In the 1770's Borda made improved versions of this instrument with the help of Lenoir and obtained excellent results. According to Delambre and Lalande, it was Borda who influenced the early committee on the metric system to decide on a measure of the meridian as the basis for a standard and to use his instrument for this purpose to establish further its reputation.

Included in the following excerpt of this chapter are interesting comments on the possible extension of decimalization to time and the compass.[3] There follow Bugge's criticisms (or rather reservations) on the determination of the metric standards. Finally he gives a short description of the method used to determine the kilogram.

Bugge discusses the practical difficulties for an instrument maker of dividing a graduated angular scale if the proposed Revolutionary right angle of a hundred degrees were introduced. To maintain decimal consistency it had been proposed that each degree should consist of a hundred minutes and each minute of a hundred seconds of arc. The author continues:

> If the new French metric system is to be introduced everywhere—which is hardly possible, because of innumerable difficulties—then all scientific connections between our descendants and our forefathers would also be severed. Just consider; suppose that posterity, using the new republican calendar, the instruments and tables of the one hundred degrees scale, only dividing the day into twelve hours, were to read the astronomical, geographical and nautical observations of to-day or earlier times, they would understand neither us nor our forebears unless they constantly translated our language into theirs and reduced the previous scales of time and space to their scales. They would soon, however, tire of these frequent calculations and reductions, cast aside the old fashioned rubbish and make little or no use of the works of their predecessors, which would be so difficult to understand, and this would apply not only to mathematical writings, but also to those in physics and chemistry involving weights and measures. This would be a great loss for the expansion of human knowledge. Advances in astronomy, geography, and hydrography rely mainly on comparisons between new and older observations, experiments

[3]The decimalization of the compass became permanent in France and anyone using French maps today should be prepared not only for decimal degrees but also a decimal longitude based on the Paris meridian.

and notes, and this will be rendered even more difficult by changing the accepted system of measurement.

In one of my last meetings with the famous Borda, he told me that he was thinking of dividing the mariners' compass according to the decimal scale. Up till now the dial of the compass has been divided into thirty-two graduations or winds, which have been given such convenient, well-chosen and suitable names, that in my opinion it would be difficult to invent better. Borda wanted to divide the compass into forty winds or graduations and, had he lived longer, it would not have been difficult for him to obtain a decree from the legislative authorities, whereby this new compass would have been introduced into the French navy and into navigational manuals. Whether this division (of the compass) would really procure greater advantages and accuracy in setting course or in discovering the longitude and latitude, I shall leave to the judgement of informed mariners. Lévèque, the gifted professor of hydrography, who has published some excellent works on navigation, denies it and maintains that on the contrary unnecessary confusion is more likely to result from it. When Borda died presumably the execution of this project died too and French sailors will probably, like sailors of all other nations, keep to the old division of the compass.

According to the new metric system the day is to be divided into ten hours, the hour into a hundred minutes and the minute into a hundred seconds. Two skilled clockmakers, Berthoud and Breguet, and several others have handed a very well drafted document to the Directory on behalf of all the clockmakers in Paris, in which the many difficulties for the clock industry and the impracticability of the new division of time were indicated. The result was that the legislative authority decided to suspend the introduction of the new republican time scale for the present. I have seen only two clocks in Paris with dials divided according to the new time.[4] One of them was set up on the facade of the Palace of the Tuileries overlooking the garden.

[4]One of these decimal clocks was seen by Bugge at the exhibition of French industry (p. 133).

It was considered a new-fangled curiosity. All the other clocks in the Tuileries and elsewhere in Paris show the old division of time and according to it people everywhere in Paris get up and go about their work until six o'clock, and they eat, go to the theatre, go to parties and to bed. The other clock that I saw with the new time scale was a nautical pocket-watch or chronometer. Berthoud had made it and it belonged to Borda; in the National Observatory Bouvard showed me the observations and record of its performance over a period of fifteen months, which was excellent.

Some difficulties may also occur as a result of the new Republican calendar, particularly in connection with the beginning of the French year, which should be the day when, according to the astronomical calculations, the sun enters the beginning or the zero-point of Libra. The sun's entrance into Libra does not always occur on the same day and in the same hour, and is bound to no fixed and unalterable time. The beginning of the Republican year therefore is also subject to change and it will be very difficult to calculate the beginning of the French year for past as well as for future time. According to the calendar year of all the other European nations, which has been approved by legislators and astronomers for more than a thousand years and found correct by long experience, it is possible to determine civil chronology for many centuries of past as well as future time by a very easy and simple intercalation of one day; only after 3,200 years will this reckoning of time diverge one day from the true astronomical calculation. Therefore the advantage gained seems very doubtful, if such an extremely simple, easy and reliable system of chronology is to be abolished. . . .

It is clear to everyone that it will be very difficult to teach the craftsmen, artists, trades people, and country people of a whole nation this new language. Although the Directory has taken a great deal of trouble introducing the new metric system and at great expense has had printed explanations distributed to the departments, yet outside Paris it is very little known and used; only the public officials of the Republic use it, because they are compelled to by orders. Most of the people living in France do not even know the new terms and much less understand them. They continue using the old weights and measures and it will be centuries before it will be possible to introduce everywhere the new weights and measures which differ so greatly in name and scale from the former.

If all states had the same weights and measures and money, this would undeniably be very convenient, particularly for those who trade with foreign nations and for scholars who wanted to make use of the studies and observations of other nations. For the rest of the nation, however, it would be immaterial. Changing the metric system of a country entails so many difficulties and such considerable expense that it seems that uniformity of weights and measures will be counted amongst the pious wishes which are never fulfilled and it would be even more difficult to call in all the coins, melt them down, and mint them anew. The states that already have a uniform system of weights and measures have no sufficient reason for adopting the new metric system. . . .

My misgivings about the precise determination of the metre are as follows:

1. Should not the base-lines at Melun and Perpignan have been measured twice? I do not doubt Delambre's accuracy at all; I know that a small part of the base-line of 140 *toises* has been measured twice; I know that the base-line at Perpignan calculated by triangulation agreed with the actual measurement to within a foot, which in a series of triangles of over 350,000 *toises* represents inevitably a very high and admirable degree of accuracy. On the other hand, however the sceptic may object that a mistake of one second in each angle of these ninety triangles would produce an uncertainty of eight to ten feet in the base-line at Perpignan. In all the other measurements of the meridian the base-lines have always been measured twice and it seems that this final stamp of authority is lacking in these measurements which are so excellent in all other respects.

2. It must be admitted that Borda's repeating circle has many great advantages. With Méchain I made a few observations with it myself and the accuracy of this instrument exceeded my

greatest expectations; despite all these advantages, however, anyone who is familiar with this instrument must nevertheless admit that the determination of the polar altitudes is far more difficult with it than their determination by observation of a horizontal angle. Borda's repeating circles, which were used to determine the latitudes of Dunkirk, Paris, Evaux, Carcassonne, and Montjouy, are fifteen to twenty inches in diameter and, although after three to six hundred observations the angle can be found to the exact second, yet I cannot help thinking that the amplitudes of the meridian arcs, which have been determined by sectors of twelve-foot radius, on whose limbs the divisions can be read directly to a single minute by means of the Nonius scale,[5] are just as reliable as the sizes or lengths of the meridian arc determined by a twenty-inch diameter Borda circle, because what can be found directly and seen by the eye is just as reliable as what has to be inferred, no matter how lucky one's estimate may be.

3. It is a great proof of the excellence of Borda's repeating circle that in a series of ninety triangles the correction for the error of all three angles amounts to only one to five seconds. Whether the whole correction falls on one angle, or on all, and in what proportion it has to be divided, is completely arbitrary. These arbitrary corrections, however, will always contribute something towards altering the length of the sides and the arcs of the meridian; the latter thus become larger or smaller than if these corrections had been produced in a different way. I am convinced that these degrees of the meridian have been measured throughout France with all the accuracy possible to human diligence and acumen, and that they exceed all previous measurements in reliability; nevertheless, there is still always an uncertainty of a few *toises* in the length of the arc of the meridian. Another factor still to be considered is that these measurements prove that the decrease in the degree of the meridian is non-uniform and deviates from the elliptical arc of the meridian. These arbitrary corrections, these small uncertainties in the

[5]See p. 144*n*.

length of the meridian arc, these small irregularities in the degrees, the working hypothesis of elliptical meridians, have the result that it cannot be assumed that the determination of the metre is so absolutely and completely borrowed from nature that there is no doubt at all about its final accuracy; as similarly there would also be some variation if measurements were to be taken in other meridians of the same or different latitudes. However often these measurements are repeated, there would always be a few *toises*, for the mathematical certainty of which no mortal man could vouch. Despite these considerations I still believe that we are so close to measuring the true length of the metre that no human skill could come closer and that we must accept $443\frac{296}{1000}$ lines of Bouguer's *toise* for the true length of the metre at $17\frac{6}{10}$ degrees on the centrigrade thermometer. This metre could be called the scientific metre in order to distinguish it from the common metre, which according to the decrees of 1 August 1793 and 7 April 1795 is 3 feet $11\frac{44}{100}$ lines or $443\frac{44}{100}$ lines. In one of my previous letters I have stated that this common metre has been accepted by the Department of Justice (Bureau of Weights and Measures) and that according to it (at least until June 1799) all the models of metres sent as originals to the departments have been constructed. Therefore in France there are two metres of different sizes, the scientific and the common. The difference between the two is $\frac{144}{1000}$ of a line or nearly $1\frac{1}{2}$ tenths of a line.

In trade and commerce, dealing with smaller lengths, this difference is trifling, but in the case of greater lengths it is certainly noticeable; for example, in a hundred metres or approximately 308 feet it is one inch $2\frac{4}{10}$ lines, and in a thousand metres it is a foot.

My next subject is the kilogram, which should be equivalent to the weight of a cubic decimetre of distilled water. For this purpose a special commission was nominated consisting of Lefèvre-Gineau and Fabbroni, joined later by Swinden and Trallès. In my previous letter I described the brass cylinder, designed to be used for this purpose. This contains eleven cubic decimetres and the kilogram should be determined all the more exactly as it is only an eleventh part of the weight of this cylinder.... I have also described the instrument above and the ingenious way in which Lefèvre-Gineau measured the height and the diameter of this cylinder.[6] Fortin has constructed a balance of an improved type as well as the required weights with their decimal subdivisions. It was so delicate and sensitive, that, if two pounds lay in each scale-pan, $\frac{1}{50}$ of a grain turned the scale.

The cylinder was measured fifty-three times in the air with the greatest possible accuracy. There were two reasons why its weight in the air was the same as that which it would have been in a vacuum: namely, because the weights as well as the cylinder were of brass and of an equal volume and because a small opening or tube was introduced into the cylinder whereby the air outside could mingle with that inside.

Now the cylinder had yet to be weighed in distilled water. This was also done with the greatest accuracy and the following reductions were made in doing so, which show with what great care Lefèvre-Gineau thought out the smallest details which could affect delicate measurements of this kind. The first reduction is due to the fact that the air acts on the counterweight, but not on the cylinder which has been lowered into the water. If this weighing were to take place in a vacuum, the balance would be raised and the counterbalance would be too heavy by a quantity of air of the same volume as the counterweight, and this therefore had to be subtracted. The second reduction is due to the fact that in water not only the

[6]Not content with simply measuring the over-all dimensions of the cylinder, Lefèvre-Gineau measured the height exactly at thirty-six regularly spaced points on a series of diameters. The diameter was also measured several times on different cross sections of the cylinder. These measurements revealed that the cylinder was in fact a very slightly truncated cone. Fifteen special measuring rods were then constructed almost exactly equal to the height of the cylinder and fifteen more equal to the diameter. The precise variation of each rod from the respective height or diameter was then recorded. Finally, to give an absolute measure of the dimensions in meters, the rods were placed end to end on the comparative scale used for the measurement of standards of length.

weight of the cylinder itself is discovered, but also the weight of the air which is in the cavity of the cylinder; this must therefore also be deducted. The third reduction is as follows: as heat expands water, like all other bodies, the former Academy of Sciences established in their experiments on the new weights and measures, that the kilogram was to be determined at a certain temperature, and zero on Réaumur's thermometer or the temperature of snow or melted ice was chosen; and so Lefèvre-Gineau and Fabbroni surrounded the vessel of distilled water, in which the cylinder was to be weighed, with ice. By doing this however they could not produce the coldness of ice in the distilled water, the mean temperature of which remained $\frac{3}{10}°$ on the Centigrade thermometer. Some experiments performed by De Luc seem to prove that water is heaviest or has the greatest density not at zero but at 5°. In collaboration with Fabbroni and Trallès, Lefèvre-Gineau investigated this matter thoroughly by the most delicate experiments and found that a body weighed in water loses more and more of its weight as the water cools, until the water reaches 4° on the Centigrade thermometer; if it becomes any cooler, however, the body loses less of its weight; the water has therefore expanded from 4° to the freezing point and its greatest density is at 4° and this density, as proved by many accurate experiments, always remains the same. Therefore the commission also decided to include the third reduction, whereby the weight of a quantity of distilled water of the same volume as the cylinder, determined at $\frac{3}{10}°$, was reduced to that weight which it must have at 4°.

It was also very important to find the true volume or the cubic content of the cylinder. In the air at a temperature of $17\frac{1}{4}°$ it was discovered to be $11\frac{20}{100}$ decimetres. Afterwards the cylinder was weighed in water of a temperature of only $\frac{3}{10}°$ and, because of the contraction of the metals at the lower temperature, its cubic content had also diminished and this diminution was determined after experiments on the expansion of brass by heat had been carried out. Finally, one more correction had to be made. When the cylinder was weighed in distilled water at $\frac{3}{10}°$, it was suspended from the arm of the balance by a brass rod, part of which was in the water and the cylinder lost more than it should in the water, namely the weight of a quantity of water of the same cubic content as the part of the rod, which was submerged. When the two last mentioned corrections had been deducted, the result was that the volume or cubic content of the cylinder was $11\frac{98}{100}$ cubic decimetres.

The final result of the work of this second special commission was: the true kilogram or the weight of a cubic decimetre of distilled water at its greatest density, or 4 degrees, is 18,827 grains or 2 pounds, 5 gros, 35 grains of the former French pound (*poids de marc*).

This second commission made Fortin, a very skilled craftsman, prepare a kilogram of platinum and various others of brass and standardise them with the greatest possible accuracy. The kilogram of platinum cannot have the same weight in air as the kilogram of brass, which actually must be used for weighing in air. Both ought to have the same weight as the weight of a cubic decimetre of distilled water weighed in a vacuum at the greatest degree of density. Suppose that a brass kilogram of about 6 cubic inches in a vacuum has the same weight as a kilogram of platinum of about $2\frac{4}{10}$ cubic inches and that when air is let into the vessel, the brass kilogram loses as much of its weight as 6 cubic inches of air weigh and the platinum kilogram on the other hand only as much as $2\frac{4}{10}$ cubic inches of air weigh, then the latter will have excess weight in air and appear to be heavier by the weight of $3\frac{6}{10}$ cubic inches of air or approximately $1\frac{1}{2}$ grains.

By his experiments Lefèvre-Gineau has also determined the weight of a French cubic foot of distilled water in French pounds:

at 4° —70 pounds 223 grains
at $\frac{3}{10}^{\circ}$ —70 pounds 141 grains
at 0° —70 pounds 130 grains

A cubic foot of distilled water is therefore ninety-three grains heavier at its greatest density than at freezing-point.

By the laws of 1 August 1793 and 7 April 1795 the kilogram has been accepted in accordance with Lavoisier's and Haüy's experiments and weighs 18,841 grains of the former French pound (*poids de marc*). This definition was adopted at the Standards Office, and according to it models were constructed for the departments. Therefore in France there are two kilograms, the general or provisional one and the scientific or the true kilogram. The difference between the two is only fourteen grains—not considerable when dealing with large masses. When hereafter French observations, experiments, measurements, and reports are read, it could be uncertain whether the provisional or general or whether the true and scientific metre and kilogram are meant. . . .

GLOSSARY

Some important terms relating to Revolutionary France

CALENDAR

The Revolutionary Calendar was introduced in France in October 1793. It was intended to eradicate the Christian associations of the Gregorian calendar with its Sundays and Saint's days. The year was to begin on 22 September, this conveniently being both the day of the autumnal equinox and the day in 1792 when the Republic was proclaimed. Thus the REPUBLICAN YEAR II began on 22 September 1793 and the year VII, for example, corresponds to September 1798—September 1799 in the Gregorian calendar. The year was divided into twelve months of thirty days each. The remaining five days of the 365 were called *sans-culottides* (translated in Bugge's account as "complementary days") and set aside for festivals and so forth.

The months were given names related to some seasonal event: *vendémiaire* (vintage: Sept.–Oct.), *brumaire* (fog: Oct.–Nov.), *frimaire* (frost: Nov.–Dec.), *nivôse* (snow: Dec.–Jan.), *pluviôse* (rain: Jan.–Feb.), *ventôse* (windy: Feb.–Mar.), *germinal* (buds: March–April), *floréal* (flowers: Apr.–May), *prairial* (meadows: May–June), *messidor* (reaping: June–July), *thermidor* (heat: July–Aug.), *fructidor* (fruit: Aug.–Sept.).

The month was divided into three weeks of ten days, called a DECADE, and each day was given a name recalling its numerical order: *primedi, duodi,* and so forth, ending with *décadi*, the day of rest. Under Napoleon the existence of the Revolutionary Calendar became something of an embarrassment and France officially returned to the Gregorian calendar from 1 January 1806.

CONVENTION — The National Convention was the constitutional and legislative assembly, elected in September 1792 and remaining in existence for three years. One of its first acts was to abolish the powers of the king. Later, faced with a crisis in the Revolutionary wars, the Convention appointed a committee of public safety to act as the executive authority. A variety of other committees was used by the Convention for legislative and administrative purposes. The most important of these from the point of view of this book was the Comité de l'instruction publique, which was responsible for founding the new system of state education.

DECADE — See CALENDAR.

DEPARTMENT — A territorial division of France, comparable to an English county. Before the Revolution the local unit had been the province, but this was associated with *ancien régime* and it was proposed that the thirty-four provinces be replaced by eighty-three departments (this number increased as the victorious French armies pushed back the frontier). Departments were intended to be comparable in size and population and each was administered by an official, the prefect, appointed by the government in Paris.

DIRECTORY — The name given to the executive power in France between August 1795 and November 1799. The five heads of state or Directors were chosen by an upper house of 250 elected members, the Council of Ancients, this in turn being supported by another chamber, the Council of Five Hundred. Although several of the Directors were personally corrupt and provincial

	administration was chaotic, it was under the Directory that France gained its first important victories in the Revolutionary wars.
TOISE	A unit of measurement under the *ancien régime*, corresponding to approximately two yards.
YEAR, REPUBLICAN	See CALENDAR.

BIBLIOGRAPHY

Visitors to France in the Post-Revolutionary Period

A Bibliographies of visitors' accounts:

Lacombe, P., *Bibliographie parisienne. Tableaux de moeurs, 1600–1880*, Paris, 1887, pp. 59–75.

Maxwell, Constantina, *The English Traveller in France, 1698–1815*, London, 1935, pp. 279–291.

Bibliography of descriptions of various institutions in Paris:

Tourneux, M., *Bibliographie de l'histoire de Paris pendant la Révolution française*, Paris, 1890–1913, vol. III, part IV ("Histoire des lettres, des sciences et des arts"), pp. 533 ff.

B Bibliographical details of previous editions of Bugge's account:

Thomas Bugge's Reise til Paris i Aarene 1798 og 1799, Copenhagen, F. Brummer, 1800, 8vo, pp. xiv + 654, 4 plates.

Thomas Bugge's Justizrathes und Professors, Reise nach Paris in den Jahren 1798 und 1799, aus dem Daenischen übersetzt, von Johann Nicolaus Tilemann, Copenhagen, F. Brummer, 1801, small 8vo, pp. viii + 718, 4 plates.

Travels in the French Republic: Containing a circumstantial view of the present state of learning, the arts, manufactures, learned societies, manners, &c. in that country, by Thomas Bÿggé [sic], Professor of mathematical astronomy in the University of Copenhagen, late commissioner from Denmark to the National Institute, and member of several learned societies and academies of sciences, translated from the

Danish by John Jones, LL.D., London, R. Phillips, 1801, 12mo, pp. xii + 405, plate.

C Books singled out for special mention because of the relevance of at least part of their contents to the *scientific* scene in France after the Revolution:

Behn, C. H., *Erinnerungen an Paris zunaechst für Aerzte geschrieben*, Berlin & Stettin, 1799.
Georg Heinrich Behn was in France (chiefly Paris) from October 1797 to May 1798. As the title suggests, he was interested particularly in hospitals but he also described the Ecole de médecine and the Lycée républican. He announced a second volume which was to include a chapter on scientific societies.

Benzenberg, J. F., *Briefe geschrieben auf einer Reise nach Paris*, 2 vols., Dortmund, 1805–1806.
Johann Friedrich Benzenberg, who had studied physics and mathematics at Göttingen, took full advantage in 1804 of the facilities in Paris for higher education in science. He regularly attended lectures at the Muséum d'histoire naturelle; he also describes some meetings of the Institute.

[Blagdon, F. W.], *Paris as it was, and as it is; or a sketch of the French Capital, illustrative of the effects of the Revolution, in a series of letters, written by an English traveller during the years 1801–1802*, 2 vols., London, 1803.
Francis William Blagdon was a London journalist. Vol. II contains some account of scientific institutions: Letter XLV, the Institute; XLIX, scientific societies; LI, education; LVII, the Collège de France; LXI, schools for public services; LXVI, the Bureau des Longitudes and the metric system; LXXIV, the Muséum d'histoire naturelle; LXXXIII, the Conservatoire des arts et métiers.

Brugnatelli, L., "Diario di Luigi Brugnatelli dal 1 Settembre al 4 Dicembre 1801," *Epistolario di Alessandro Volta*, IV, Bologna, 1953, 461–533.
Luigi Brugnatelli accompanied Volta when he visited Paris to demonstrate his pile to the Institute. Brugnatelli's diary provides a valuable commentary on scientific theories of the time. He also describes meetings with the leading

French men of science and mentions the various institutions he visited.

Eyre, E. J., *Observations made at Paris during the Peace*, Bath, 1803.

The dramatist Edmund John Eyre visited Paris in 1802. His book includes a short account of the Paris Observatory and of some of the educational institutions, including the Ecole de médecine.

Faber, G. T. von, *Sketches of the Internal State of France*, 2nd ed., London, 1813.

Gotthilf Theodor von Faber, a German idealist, went to France after the Revolution to offer his services. He held a post in the civil administration until 1807. Disillusioned, he then went to St. Petersburg and published volume I of a two-volume work. The second volume was suppressed through the influence of Napoleon, who also succeeded in impounding most of the copies of volume I. Chapter 5 of the English edition (pp. 140–164) deals with public instruction.

Fischer, G., *Das Nationalmuseum der Naturgeschichte zu Paris*, Frankfurt am Main, 1802.

Gotthelf Fischer's detailed guide to the Muséum d'histoire naturelle, written for German-speaking visitors to Paris, deserves to be better known.

Frank, J., *Reise nach Paris, London, und einem grossen Theile des übrigen Englands und Schottlands in Beziehung auf Spitäler, Versorgungshäuser, übrige Armen Institute, Medizinische Lehranstalten und Gefängnisse*, 2 vols., Vienna, 1804–1805.

Dr. Joseph Frank taught pathology and general therapy at the Russian University of Vilna. Although most of his account deals with Britain, a general description of the medical and scientific institutions of Paris is given in vol. 1, pp. 1–174.

Heinzmann, Johann Georg, *Voyage d'un allemand à Paris et son retour par la Suisse*, Lausanne, an VIII [1800].

This is a French translation of the German edition (*Meine Frühstunden in Paris*, 2 vols., Basel, 1800). Johann Georg Heinzmann, a bookseller from Ulm, was in Paris in 1798. Among the wide variety of subjects he discusses, there is a short section devoted to "Le monde savant et l'instruction publique" (pp. 110–117).

Kotzebue, A. F. F. von, *Travels from Berlin through Switzerland to Paris in the year 1804*, 3 vols., London, 1804.
This is a translation from the German edition (*Erinnerungen aus Paris im Jahre 1804*, Berlin, 1804). The dramatist, August Friedrich Ferdinand von Kotzebue, includes some account of the arts and sciences in vols. II and III of the English edition.

Meyer, F. J. L., *Fragments sur Paris, traduits de l'allemand par le général Dumouriez*, 2 vols., Paris, 1798.
This is a translation from the German edition (*Fragmente aus Paris im IVten Jahr der französischen Republik*, 2 vols., Hamburg, 1797). Friedrich Johann Lorenz Meyer, a lawyer from Hamburg, stayed in Paris from 31 March to 4 July 1796. Vol. II mentions the Institute, the Jardin des plantes, the Observatory, the École polytechnique, as well as ballooning, telegraphs, and factories.

Pictet, M. A., "Journal d'un Genevois à Paris sous le Consulat," *Mémoires et documents publiés par la société d'histoire et d'archéologie de Genève*, 2e série, V (1893–1901), 98–133.
Marc Auguste Pictet from Geneva was one of the editors of the *Bibliothèque britannique*. He was in Paris in 1801 and again in 1802 and kept in close contact with the leading scientists in Paris.

Pinkerton, J., *Recollections of Paris in the years 1802–1803–1804–1805*, 2 vols., London, 1806.
The Scottish antiquarian and historian, John Pinkerton, was the author of several travel books. Having stayed in France in 1803 after the rupture of the Peace of Amiens, he was imprisoned but was released in 1805 through the intercession of Sir Joseph Banks, president of the Royal Society. Vol. I contains the inevitable chapter on the *Jardin des plantes* and chap. 16 is concerned with learned societies. Vol. II, chap. 20, lists literary and scientific journals.

Pujoulx, J. B., *Promenades au Jardin des Plantes, à la Ménagerie et dans les Galeries du Muséum d'Histoire Naturelle. Contenant des Notions claires et à la portée des Gens du monde sur les végétaux, les animaux et les minéraux les plus curieux et les plus utiles de cet Établissement. Ouvrage principalement*

destiné aux Personnes qui le visitent, 2 vols., Paris, an XII [1804].
Tome Premier. Jardins, Serres, etc., Parcs, Ménageries, Salle du Règne Végétal, Galeries des Minéraux, Salle des Fossiles. Avec des Additions relatives aux accroissements de la Ménagerie.
Tome Second. Galerie et Salle du Second Etage, Contenant les Diverses Classes d'Animaux. Notices sur les distributions gratuites des plantes, la collection d'anatomie, la bibliothèque, les cours publics, etc.

Reichardt, J. F., *Un hiver à Paris sous le Consulat, 1802–1803, d'après les lettres de J. F. Reichardt,* A. Laquiante, ed., Paris, 1896.
This is based on Reichardt's *Vertraute Briefe aus Paris* (3 vols., Hamburg, 1804), of which there was a French translation published at Fribourg, 1858. Johann Friedrich Reichardt gives some thumbnail sketches of men of science, but he was really more interested in the theater.

Shepherd, W., *Paris in 1802 and 1814,* London, 1814.
Rev. William Shepherd was one of the many British visitors to Paris during the Peace of Amiens. His account includes a description of the Institute (pp. 100–102).

Yorke, H. R., *Letters from France; describing the manners and customs of its inhabitants; with observations on the arts and manufactures, public institutions and buildings, learned societies, and the mode of travelling. Interspersed with interesting anecdotes of celebrated public characters,* 2 vols., London, 1814 (written in 1802).
Henry Redhead Yorke's account contains many observations on scientific institutions. The style, however, is sometimes intemperate, reflecting Yorke's violent change of political opinion.

General Bibliography

Artz, F. B., *The Development of Technical Education in France, 1500–1850,* Cambridge, Mass., and London, 1966.

Biot, J. B., *Essai sur l'histoire générale des sciences pendant la Révolution française,* Paris, 1803.

Bugge, M., "Grundzüge zu Thomas Bugge's Lebensbeschreibung," *Zeitschrift für Astronomie*, II (1816), 245–250.

Crosland, M. P., *The Society of Arcueil. A View of French Science at the Time of Napoleon I*, Cambridge, Mass., and London, 1967.

Daumas, M. (ed.), *L'expansion du machinisme. Histoire générale des techniques*, vol. III, Paris, 1968.

Décade philosophique, La, Paris, 1794–1804.

Fayet, J., *La Révolution française et la science, 1789–1795*, Paris, 1960.

Gillispie, C. C., "Science and Technology." In *New Cambridge Modern History*, vol. V, edited by C. W. Crawley, Cambridge, 1965.

Merz, J. T., *A History of European Thought in the Nineteenth Century*, vol. I, part 1 ("Scientific Thought"), London, 1905, reprint, New York, 1965.

Smeaton, W. A., *Fourcroy, Chemist and Revolutionary: 1755–1809*. Cambridge, England, 1962.

Stein, J. W., *The Mind and the Sword*, New York, 1961.

Taton, R. (ed.), *Enseignement et diffusion des sciences en France au XVIIIe siècle*, Paris, 1964.

CHAPTER 1

Ackerknecht, E. H., *Medicine at the Paris Hospitals, 1794–1848*, Baltimore, 1967.

Aguillon, L., "L'École des mines de Paris. Notice Historique," *Annales des mines, 8e série, Mémoires*, XV (1889), 433–686.

Denise, L., *Bibliographie historique et iconographique du Jardin des plantes*, Paris, 1903

Dupuy, P., *L'Ecole normale de l'an III*, Paris, 1895.

École polytechnique, *Livre du Centenaire, 1794–1894*, vol. I (*L'École et la Science*), Paris, 1895.

Fourcy, A., *Histoire de l'École polytechnique*, Paris, 1828.

Lacroix, S. F., *Essai sur l'enseignement en général et sur celui des mathématiques en particulier*, Paris, 1805.

Lefranc, A., *Histoire du Collège de France*, Paris, 1893.

Liard, L., *L'Enseignement supérieur en France, 1789–1889*, 2 vols., Paris, 888, 1894.

Prevost, A., *La Faculté de Médecine de Paris. Ses chaires, ses annexes et son personnel enseignant de 1794 à 1900*, Paris, 1900.

Raymond-Latour, J. M., *Souvenirs d'un Oisif*, 2 vols., Lyons and Paris, 1836.

Williams, L. P., "Science, Education and the French Revolution," *Isis*, XLIV (1953), 311–330.

CHAPTER 2

Aucoc, L., *L'Institut de France. Lois, Status et Règlements concernant les anciennes Académies et l'Institut de 1635 à 1889*, Paris, 1889.

Gauja, P., *L'Académie des Sciences de l'Institut de France*, Paris, 1934.

Institut de France, *Procès-verbaux des séances de l'Académie des Sciences, tenues depuis la fondation de l'Institut jusqu'au mois d'août 1835, publiés conformement à une decision de l'Académie par MM. les Secrétaires perpétuels*, 10 vols., Hendaye, 1910–1922.

Maindron, E., *L'Académie des Sciences*, Paris, 1888.

Simon, J., *Une Académie sous le Directoire*, 1885.

CHAPTER 3

Bigourdan, G., "Le Bureau des Longitudes. Son histoire et ses travaux de l'origine (1795) à ce jour," *Annuaire du Bureau des Longitudes*, 1928, A1–A72; 1929, C1–C92; 1930, A1–A110; 1931, A1–A145; 1932, A1–A117.

Cassini, J. D., *Mémoires pour servir à l'histoire des sciences et à celle de l'Observatoire Royal de Paris*, Paris, 1810.

Devic, J. F. S., *Histoire de la vie et des travaux de J. D. Cassini IV*, Clermont (Oise), 1851.

Lalande, J. le Français de, *Bibliographie astronomique avec l'histoire de l'astronomie depuis 1781 jusqu'à 1802*, Paris, 1803.

Wolf, A., "Astronomical Instruments," *A History of Science, Technology and Philosophy in the Eighteenth Century*, chap. 5, 2d ed., London, 1952.

Wolf, C., *Histoire de l'observatoire de Paris de sa fondation à 1793*, Paris, 1902.

CHAPTER 4

Cardwell, D. S. L., *Steam Power in the Eighteenth Century*, London, 1963.

Colmont, A. de, *Histoire des expositions des produits de l'industrie française*, Paris, 1855.

Levasseur, E., *Histoire des classes ouvrières et de l'industrie en France depuis 1789*, 2 vols., Paris, 1903.

Payen, J. "Bétancourt et l'introduction en France de la machine à vapeur à double effet (1789)," *Revue d'histoire des sciences*, XX (1967), 187–198.

Prony, R., *Nouvelle architecture hydraulique. Seconde Partie, contenant la description détaillée des machines à feu*, Paris, 1796.

Richard, C., *Le comité de salut public et les fabrications de guerre sous la Terreur*, Paris, 1922.

CHAPTER 5

Berthelot, P. E. M., "Sur les publications de la Société Philomatique et sur ses origines," *Journal des Savans*, August 1888, pp. 477–493.

Bulletin des Lois, , 2me série, VII, no. 239.

Charmes, X., "Bibliographie des sociétés savantes de la France," *Collection de documents inédits sur l'histoire de la France*, II, Paris, 1886, pp. 475–586.

Daumas, M., *Les instruments scientifiques au XVIIe et XVIIIe siècles*, Paris, 1953.

Smeaton, W. A., "The early years of the *Lycée* and the *Lycée des Arts*. A chapter in the lives of A. L. Lavoisier, and A. F. de Fourcroy," *Annals of Science*, XI (1954), 257–267, 309–319.

CHAPTER 6

Babeau, A., *Le Louvre et son Histoire*, Paris, 1895.

Bazin, G., *The Louvre*, trans., London, 1957.

Dowd, D. L., *Pageant-Master of the Republic. Jacques-Louis David and the French Revolution*, University of Nebraska Studies, New Series, No. 3, Lincoln, Nebraska, June 1948.

Saunier, C., *Les conquètes artistiques de la Révolution*, Paris, 1902.

CHAPTER 7

Bigourdan, G., *Le système metrique des poids et mesures*, Paris, 1901.

Crosland, M.P., "The Paris Congress on Definitive Metric Standards (1798–1799)—The First International Scientific Congress?" *Isis,* LX (1969), 226–231.

Favre, A., *Les origines du système metrique*, Paris, 1931.

Méchain, P. F. A. and Delambre, J. B. J., *Base du système metrique decimal ou mesure de l'arc du méridien compris entre les parallèles de Dunkerque et Barcelona, Suite des Mémoires de l'Institut*, 3 vols., January 1806–November 1810.

Swinden, J. H. van, *Verhandeling over Volmaakte Maaten en Gewigten*, 2 parts, Amsterdam, 1802 (not seen).

NAME INDEX

PLACE INDEX

SUBJECT INDEX